Francis Ellingwood Abbot

Scientific Theism

Francis Ellingwood Abbot

Scientific Theism

ISBN/EAN: 9783337415372

Printed in Europe, USA, Canada, Australia, Japan

Cover: Foto ©Lupo / pixelio.de

More available books at **www.hansebooks.com**

ORGANIC SCIENTIFIC PHILOSOPHY

SCIENTIFIC THEISM

BY

FRANCIS ELLINGWOOD ABBOT, Ph.D.

BOSTON
LITTLE, BROWN, AND COMPANY
1885

To

The Hallowed Memory

of

My Mother.

———•———

Tender and true, with love's deep wisdom wise,
    Thou taught'st the Child the free, pure Truth to crave,
More than earth's gold the gold of God to prize;
    And now the Man, who inly burned to have
Thy joy for his reward, with blinded eyes
    Lays the won gold, unheeded, on thy grave.

PREFACE.

THE foundation and immediate occasion of this little book, whose size, I trust, is no necessary measure of its usefulness, was a lecture given before the Concord Summer School of Philosophy, July 30, 1885, in a "symposium" on the question: "Is Pantheism the Legitimate Outcome of Modern Science?" The other lecturers on this subject were Mr. John Fiske, Prof. William T. Harris, Rev. Dr. Andrew P. Peabody, Prof. George H. Howison, and Dr. Edward Montgomery, — the lectures of the last two gentlemen being read by Mr. Thomas Davidson. The contents of my own lecture, entirely re-written from the first page, constitute less than one third of what is here printed.

The real origin of the book, however, was two articles published in 1864 in the *North American Review*, while it was still under the scholarly care and joint editorial management of Professors James Russell Lowell and Charles Eliot Norton, — one in the July number on "The Philosophy of Space

and Time," and the other in the October number on
" The Conditioned and the Unconditioned."

Some of the criticisms here made on Mr. Herbert
Spencer's philosophy, for much of which I have the
highest admiration, were embodied in a general article on his *First Principles*, entitled "Positivism in
Theology," and published in the now discontinued
Christian Examiner in Boston, March, 1866 ; and in
a special and elaborate review of his *Principles of
Biology*, published in the *North American Review*
for October, 1868, under the caption "Philosophical
Biology." To both of these articles Mr. Spencer
made replies, which to my mind were eminently inadequate and unsuccessful, — to the former, through
Prof. E. L. Youmans, in a subsequent issue of the
Christian Examiner, and to the latter in a special
pamphlet, entitled *Spontaneous Generation*, and published by D. Appleton and Company in 1870. I
make these references in fairness to Mr. Spencer,
that those who wish may investigate the subject
more fully.

The theory of Phenomenism *versus* the theory
of Noumenism ; the theory of Idealistic Evolution
versus the theory of Realistic Evolution ; and the
Mechanical theory of Realistic Evolution *versus* the
Organic theory of Realistic Evolution, — these are
the vital philosophical problems of our century, and
their solution must determine and decide that of the
vital religious problem of Theism, Atheism, and
Pantheism. The discussion of these problems con-

stitutes the substance of this book; and I must express my belief (not, I trust, without becoming modesty, for I submit my own belief unreservedly to the final verdict of the universal reason of mankind) that it formulates a philosophical revolution, since it substitutes the philosophized scientific method for the now accepted phenomenistic method, in the settlement of all philosophical questions. In the opening lecture of the "symposium" above mentioned, Mr. Fiske referred to the "revolution effected by the influence of modern science upon modern philosophy" (I quote from memory only), but did not show what this revolution is. To show what it is, and to what it leads in the sphere of religious belief, is the special object of my book.

For a quarter of a century it has been my growing conviction that the solution of all the problems named can only be accomplished by the principle of the OBJECTIVITY OF RELATIONS, together with its correlative and derivative principle of the PERCEPTIVE UNDERSTANDING. In my article on "The Philosophy of Space and Time," published (as already stated) in the *North American Review* for July, 1864, occurs the following passage, which not obscurely hints at these two fundamental principles of a reformed modern philosophy : —

"Now the five modifications of extension above described [magnitude, form, position, distance, and direction] are all *relations* among the limits of extension ; and, inasmuch as relations cannot possibly be objects of sensuous perception, but only of a higher faculty, it follows that extension alone, and not its modifications, is immediately cognized by sense. Whether these relations can in any way be cognized immediately, or only by a process of inference, it is unnecessary here to inquire; suffice it to say that, if we really *know* the objective relations of things, there must be some faculty of pure and immediate cognition of relations."

. The novelty of this book lies in its acceptance, on the warrant of modern science and the scientific method, of the fact that we *do* "*know* the objective relations of things," and in its attempt to develop the necessary philosophical implications and consequences of this fact, which phenomenistic modern philosophy steadily denies. From 1864 to the present time, I have followed the clew of the two fundamental principles above emphasized, and have been guided by them to results which, if true, must prove to be of incalculable importance and influence, not only in philosophy, but also in religion. This thin volume was written at Nonquitt Beach in five summer weeks; but it took five times five years to think it out. It is a mere *résumé* of a small portion of a comprehensive philosophical system, so far as I have been able to work it out under most distracting, discouraging, and unpropitious circumstances of many years; and for this reason I must beg some indulgence for the unavoidable incompleteness of my work. It is not the last word I hope to say on philosophy, if this word is kindly welcomed; but that remains to be proved, and in the afternoon of life the time is growing short.

Hegel argues that, just as the other sciences start with the subjective presupposition, or postulate, of the existence of their object-matter, so it would seem that philosophy must start with the subjective pre-supposition, or postulate, of the existence of its own object-matter, *thought.* But he denies the parallelism of the two cases. He maintains that, though philosophy must start with some initial position or "immediate standpoint," this immediate standpoint must, in the course of the science, be converted into a final result; and that in this manner philosophy exhibits the form of a closed or "self-returning circle" (*ein in sich zurückgehender Kreis*), whose curve sweeps back to its starting point, and, by meeting, effaces it. "The only end, act, and aim of philosophy is to attain to the notion of its own notion, and thus to its own self-return and self-satisfaction." [1]

I might perhaps claim that, even by this Hegelian canon, Scientific Realism may be adjudged to be a true philosophy, notwithstanding Hegel's other canon that "every true philosophy is Idealism." [2] For the existence of the Real Universe, which the scientific method in its empirical use apparently presupposes

[1] *Werke,* VI. 25, 26.
[2] "Jede wahrhafte Philosophie ist deswegen Idealismus." (*Werke,* VI. 189.)

as a mere postulate, and which I adopt as my own
initial position on the warrant of the scientific
method, is at the end (§ 87) explained as a speculative final result in the Eternal Creative Act:
" The absolute ' full-filling ' of Thought-in-itself, therefore, or the embodiment of the Ideal in the Real, is
the eternal self-legislation of Thought-in-itself into
Thought-in-Being — of the subjective relational system into the objective relational system of the Real
Universe." In thus " attaining to the notion of its
own notion," my philosophy may be justly said to
constitute a closed or " self-returning circle."

But the apparent postulate of the scientific
method is no postulate, no " subjective presupposition," at all. On the contrary, the presuppositions
of the scientific method are formulated objective
perceptions; they are made on the authority of the
perceptive understanding (§ 50), which is every whit
as valid as that of the philosophic reason, is itself
" presupposed " by the latter, needs no higher sanc-
tion than itself, and at last, as the supreme organon
of Verification, summons the philosophic reason itself to its own tribunal for the judicial valuation of
its " final results." Here lies the profound difference
between scientific realism and philosophical idealism,
stated as follows in the text (§ 69): " Hegel sublimely disregards the distinction between Finite
Thought and Infinite Thought: the latter indeed
creates, while the former *finds,* its object. And,
since human philosophy is only finite, it follows

that *no* true philosophy is Idealism, except the Infinite Philosophy or Self-Thinking of God."

I call attention to these points here, that the Hegelian antipathy to "presuppositions" may not lead any of my readers, when they see that scientific theism rests ultimately on the presuppositions of the scientific method, to lay down my book in disgust. I venture to ask them to read it through to the end, and to consider thoughtfully whether there may not be truth, after all, in results which are undeniably at variance with current philosophic opinions.

In conclusion, I would say to my critics: "May you be fair and just enough to take pains to understand before you criticise! For then I shall be only too glad to profit by your criticisms." And, believing that there are innumerable minds in this age which have lost faith in the old without finding faith in the new, I would say to my readers: "May the hard-won thought of my little book be so clearly *truth* to your minds, that it may bring you renewed peace, serenity, and repose in the Infinite Soul of All!"

F. E. A.

CAMBRIDGE, MASSACHUSETTS,
    September 15, 1885.

CONTENTS.

CHAPTER II.

THE THEORY OF PHENOMENISM.

CHAPTER III.

THE THEORY OF NOUMENISM.

PART II.

THE RELIGION OF SCIENCE.

CHAPTER IV.

THE PRINCIPLES OF SCIENTIFIC THEISM.

CHAPTER V.

THE UNIVERSE: MACHINE OR ORGANISM?

CHAPTER VI.

THE GOD OF SCIENCE.

GENERAL SYNOPSIS

ARGUMENT FOR SCIENTIFIC THEISM.

———⊙———

I. The Foundation of Scientific Theism is the Philosophized Scientific Method.

II. The Ground-Principle of the Philosophized Scientific Method is the Infinite Intelligibility of the Universe *per se*.

    1. What is Intelligibility?

    Ans. Intelligibility is the Possession of an Immanent Relational Constitution.

    2. What is Intelligence?

    Ans. Intelligence is —

        (1) The Sole Discoverer of Immanent Relational Constitutions.

        (2) The Sole Creator of Immanent Relational Constitutions.

        (3) Identical in all Forms, and in all Teleological.

III. The Infinite Intelligibility of the Universe proves its
Infinite Intelligence, because only an Infinite Intelligence could create an Infinite Relational Constitution.

IV. The synchronous Infinite Intelligibility and Infinite Intelligence of the Universe prove that it is an Infinite
Subject-Object, or Infinite Self-conscious Intellect.

V. The Immanent Relational Constitution of the Universe-
Object, being Infinitely Intelligible, must be an Absolutely Perfect System of Nature : therefore —

    1. Not Chaos, which would be no System at all.
    2. Not a mere Multitude of Monads or Atoms,
       which would be an Unintelligible Aggregate of
       Systems.
    3. Not a mere Machine, which would be an Imper-
       fect System.
    4. But a Cosmical Organism, which is the only
       Absolutely Perfect System.

VI. The Infinitely Intelligible and Absolutely Perfect Organic
System of Nature proves that the Universe-Object is
the Eternal, Organic, and Teleological Self-Evolution
of the Universe-Subject — the Eternal Self-Realization
or Self-Fulfilment of Creative Thought in Created
Being — the Infinite Life of the Universe *per se.*

VII. The Infinite Organic and Organific Life of the Universe
per se proves that it is Infinite Wisdom and Infinite
Will — Infinite Beatitude and Infinite Love — Infinite
Rectitude and Infinite Holiness — Infinite Wisdom,

Goodness, and Power — Infinite Spiritual Person — the LIVING AND LIFE-GIVING GOD FROM WHOM ALL THINGS PROCEED.

VIII. Therefore, the Philosophized Scientific Method creates the only Idea of God which can at once satisfy both Head and Heart; and Scientific Theism creates the only Real Reconciliation of Science and Religion.

INTRODUCTION.[1]

I.

In the Preface to the Second Edition of the *Critique of Pure Reason*, Kant has this remarkable passage : —

"It has hitherto been assumed that our cognition must conform to the objects; but all attempts to ascertain anything about these *à priori*, by means of conceptions, and thus to extend the range of our knowledge, have been rendered abortive by this assumption. Let us, then, make the experiment whether we may not be more successful in metaphysics, if we assume that the objects must conform to our cognition. . . . We here propose to do just what Copernicus did in attempting to explain the celestial movements. When he found that he could make no progress by assuming that all the heavenly bodies revolved around the spectator, he reversed the process, and tried the experiment of assuming that the spectator revolved, while the stars remained at rest. We may make the same experiment with regard to the intuition of objects. If the intuition must conform to the nature of the

---

[1] Reprinted from the London *Mind* for October, 1882, where it appeared with the title, "Scientific Philosophy: A Theory of Human Knowledge."

1

objects, I do not see how we can know anything of them *à priori*. If, on the other hand, the object conforms to the nature of our faculty of intuition, I can then easily conceive the possibility of such an *à priori* knowledge. . . . This attempt to introduce a complete revolution in the procedure of metaphysics, after the example of the geometricians and natural philosophers, constitutes the aim of the Critique of Pure Speculative Reason."

Lange, in his *History of Materialism* (II. 156), thus alludes to the foregoing passage, and correctly states the conclusions logically deducible from it : —

"Kant himself was very far from comparing himself with Kepler; but he made another comparison that is more significant and appropriate. He compared his achievement with that of Copernicus. But this achievement consisted in this, that he reversed the previous standpoint of metaphysics. Copernicus dared, 'by a paradoxical but yet true method,' to seek the observed motions, not in the heavenly bodies, but in their observers. Not less 'paradoxical' must it appear to the sluggish mind of man, when Kant lightly and certainly *overturns our collective experience, with all the historical and exact sciences*,[1] by the simple assumption that our notions do not regulate themselves according to things, but things according to our notions. It follows immediately from this that the objects of experience altogether are only *our* objects; that the whole objective world is, in a word, not absolute objectivity, but only objectivity for man and any similarly organized beings, while, behind the phenomenal world, the absolute nature of things, the 'thing-in-itself,' is veiled in impenetrable darkness."

[1] The italics are ours.

Now when the great Kant, whose towering and consummate genius there is no one to dispute, founded the Critical Philosophy on this cardinal doctrine that "things conform to cognition, not cognition to things," and when he claimed thereby to have created a mighty "revolution" in philosophy comparable only with that of Copernicus in astronomy, did he really occupy a new philosophical standpoint, or really adopt a new philosophical method?

No. On the contrary, he merely completed, organized, and formulated the veritable revolution which was initiated in the latter half of the eleventh century by Roscellinus the Nominalist, — which was condemned in his person by the Realist Council of Soissons, revived in the fourteenth century by William of Occam, and finally made triumphant in philosophy towards the end of the fifteenth century, not so much by the inherent strength of Nominalism as by the weakness of its expiring rival, Scholastic Realism.

The essence of Nominalism was the doctrine that universals, or terms denoting genera and species, correspond to nothing really existent outside of the mind, but are either mere empty names (Extreme Nominalism) or names denoting mere subjective concepts (Moderate Nominalism or Conceptualism). Nominalism distinctly anticipated the Critical Philosophy in referring the source of all general conceptions (and thereby of all human knowledge), not to the object alone or to the object and subject together, but to the subject alone; it distinctly anticipated the doctrine that "things conform to cognition, not cognition to things." Since genera and species are classifications of things based on their supposed resemblances and differences, the denial of all objective reality to genera and species is the denial of all objective reality to the supposed resemblances and differences of things themselves ; the denial of all knowledge of the relations of objects is the denial of all knowledge of the objects related; and this denial is tantamount to the assertion that things-in-themselves are utterly unknown.

Wrapped up in the essential doctrine of Nominalism, therefore, was the doctrine that things-in-themselves are utterly unknown; that the knowledge of their supposed resemblances and differences is derived only from the supposing mind; that "things conform to cognition, not cognition to things;" in short, that the only knowledge possible to man is the knowledge of the *à priori* constitution of his own mind, and the relations which it imposes upon things (if they exist), totally irrespective of what things really are.

Nothing can be plainer, then, than that the Critical Philosophy did but logically develop the prime tenet of Nominalism, formulate it successfully, and expand it to a self-consistent philosophical system. This, and this alone, was the true merit of Kant. The "revolution" by which philosophy was made to transfer its fundamental standpoint from the world of things to the world of thought, and in consequence of which modern philosophy in both its great schools has inherited an irresistible tendency towards Idealism, had been substantially effected and definitely established some four hundred years before. Kant did but bring to flower and fruitage the seed sown by Roscellinus, and his Critical Philosophy was only the logical evolution and outcome of Mediæval Nominalism.

By Kant's masterly development of Nominalism into a great philosophical system, it has exercised upon subsequent speculation a constantly increasing power. In truth, all modern philosophy, by tacit agreement, rests upon the Nominalistic theory of universals. Hence alone can be explained the fact, so patent and so striking, yet so little understood or even inquired into, that both the great schools of modern philosophy, the Transcendental and the Associational, equally exhibit in its full force the tendency to Idealism latent in that theory. Nominalism logically reduces all experience, actual or possible, to a mere subjective affection of the individual Ego, and does not permit even the Ego to know itself as a noumenon. The historical development of the Critical Philosophy into the subjective idealism of Fichte, the objective idealism of Schelling, and the absolute idealism of Hegel, only shows how impossible it is for that philosophy to overstep the magic circle of Egoism with which Nominalism logically environed itself. No less striking is the inability of the English school to escape from the idealistic tendencies inherent in its purely subjective principle of Association —one of the innumerable *aliases* by which Nominalism eludes detection at the bar of contemporary thought; for Locke's successors, Berkeley, Hume, Hartley, the Mills, Bain, Spencer, and others, drift towards Idealism as steadily as Kant and his successors. It is, in fact, logically impossible to draw any but idealistic conclusions from the premises of Nominalism — and those, too, idealistic conclusions which cannot stop short of absolute Solipsism.

That modern philosophy in both its great branches irresistibly tends to Idealism is a position that will scarcely be disputed. Dr. Krauth, in his admirable edition of Berkeley's *Principles of Human Knowledge* (p. 122), thus sums up the grounds of this general and admitted tendency, while yet not perceiving that in the last analysis they are all reducible to the almost universal acceptance of the Nominalistic view of genera and species, with its implied negation of the objectivity of relations : —

"It [Idealism] rests on *generally* recognized principles in regard to consciousness. Its definition of consciousness is the one most widely received : the mind's recognition of its own conditions. It maintains that the cognitions of consciousness are absolute and infallible, and that nothing but these is, in their *degree*, knowledge. In all these postulates the great mass of thinkers agree with Idealism. The foundation of Idealism is the common foundation of nearly all the developed philosophical thinking of all schools. Idealism declares that, while consciousness is infallible, our interpretations of it, on which we base *inferences*, may be incorrect; and nearly all thinkers of all schools agree with Idealism here. No inference, or class of inferences, in which a mistake ever occurs is a basis of positive knowledge. Hence, says Idealism, only that which is directly in consciousness. is positively known, and nothing is directly in consciousness but the mind's own states. Therefore we *know* nothing more. So completely has this general conviction taken possession of the philosophical mind, that even antagonists of Idealism, who would cut *it* up by the roots if they could cut *this* up, have not pretended that it could be done." (The italics are all Dr. Krauth's.)

The "strength of Idealism," thus described by Dr. Krauth, is the strength of Nominalism — no more, no less. If all the general and special relations of things, conceived by the mind and expressed by general terms, exist in the mind alone, nothing is known of things themselves; for knowledge of things is knowledge of their relations. Nominalism, therefore, is the original source of the definition of knowledge adopted by Idealism, as shown above: that is, the contents of consciousness alone. Inasmuch, moreover, as the notion of a *common* consciousness is itself a general notion, and consequently destitute of all objectivity, nothing is "knowledge," so defined, that is outside of the *individual* consciousness. Beginning with Nominalism, therefore, Idealism must end in Solipsism, on penalty of stultifying itself by arbitrary self-contradiction. This was the path marked out for the Critical Philosophy by inexorable logic, and Fichte was more Kantian than Kant himself when he resolutely pursued it. Solipsism is the very *reductio ad absurdum* of Idealism, yet it is the rigorously logical consequence of its own definition of knowledge, which again is the rigorously logical consequence of the Nominalistic view of universals. On this point, a further quotation from Dr. Krauth will be extremely pertinent: —

"While Idealism has here a speculative strength, which it is not wise to ignore, it is not without its weakness, even at this very point, for its history shows that it is rarely willing to stand unreservedly by the results of its own principle as regards consciousness. If it accept only the direct and infallible knowledge supplied in consciousness, it has no common ground left but this — that there is the one train of ideas, which passes in the consciousness of a particular individual. A consistent Idealist can claim to know no more than this — that there exist ideas in his consciousness. He cannot know that he has a substantial personal existence, or that there is any other being, finite or infinite, beside himself. And as many Idealists are not satisfied with maintaining that we do not know that there is an external world, but go further, and declare that we know that there is not an external world, they must for consistency's sake hold that an Idealist knows that there is nothing, thing or person, beside himself. Solipsism, or absolute Egoism, with the exclusion of proper personality, is the logic of Idealism, if the inferential be excluded. But if *inference*, in any degree whatever, be allowed, not only would the natural logic and natural inference of most men sweep away Idealism, but its own principle of knowledge is subverted by the terms of the supposition. Idealism stands or falls by the principle that *no inference is knowledge.* We may reach inferences by knowledge, but we can never reach knowledge by inference" (p. 123).

Against both schools of modern philosophy, therefore, committed as they both are to the definition of knowledge drawn from Nominalism and ending in Solipsism, the charge of logical inconsistency and self-contradiction may be fairly brought, just so far as they hesitate to follow up the path to cloudland which begins with that definition. But any philosophy which hesitates to be logical forfeits all claim to the respectful consideration of mankind.

The great Roscellino-Kantian " revolution " by which Nominalism was made to supplant Scholastic Realism, and philosophy to transfer its fundamental standpoint from the world of things to the world of thought, was a revolution which logically contracts "human knowledge" to the petty dimensions of individual self-consciousness — renders it valueless as to things themselves and valuable only as to the *à priori* constitution of the individual's own mind — and in effect reduces it to a grand hallucination. Like the French Revolution, the Nominalistic revolution can live only by the guillotine, and decapitates every perception which pretends to bring to the miserable solipsist, shut up in the prison of his own consciousness, the slightest information as to the great outside world. Defining knowledge as the mere contents of consciousness, it relegates to non-entity, as pseudo-knowledge, whatever claims to be more than that. Under its sway, philosophy is blind to the race, and beholds the individual alone. What wonder that, in the hands of those who insist on their rights to reduce theory to practice, philosophy is so often found pandering to the moral lawlessness of an Individualism that sets mere personal opinion above the supreme ethical sanctities of the universe? In human society, individual autonomy is universal antinomy; for the law that binds only one binds none. Yet, with Nominalism for its root, Idealism for its flower, and Solipsism for its fruit, how can modern philosophy, teaching in both its great schools that the individual mind knows nothing except the states of its own consciousness, discover any law that shall have recognized authority over all consciousnesses? For such a discovery it is hopelessly incompetent. So far, therefore, as the social and moral interests of mankind are concerned, the present philosophical situation has become simply intolerable.

Fortunately for the future of society, however, the principle of cognition embodied in the Nominalistic definition of knowledge has never obtained general assent outside of the circle of purely speculative thought. The protest of "common sense" against it was even taken up by the Scotch school in the name of philosophy itself; but the same Nominalism which paralyzes all modern philosophy paralyzed the Scotch school, and the protest died on its tongue. Without any conscious protest, however, though with an instinctive hostility to "metaphysics" and to the philosophy which it confounds with "metaphysics," physical science has immovably planted itself on a new definition of knowledge, and fortified it impregnably against all comers; and, on the principle of cognition which it establishes, universal science, carrying up the physical and the mental into the higher unity of the cosmical, is even now beginning to build a temple of truth destined to be coeval with the human race.

1. Modern Philosophy defines knowledge as the recognition by the Ego of its own conscious states.

2. Modern science defines knowledge as twofold, — *individual knowledge,* or the mind's cognition of its own conscious states *plus* its cognition of the Cosmos of which it is a part, and *universal knowledge,* or the sum of all human cognitions of the Cosmos which have been substantiated by verification and certified by the unanimous consensus of the competent.

This latter definition may never have been formulated before, but it is tacitly assumed in all investigations conducted according to the scientific method; and the results of that method would be completely invalidated, if the definition itself should be essentially erroneous. Science does not present its truths as anybody's "states of consciousness," but as cosmical facts, acknowledgment of which is binding upon all sane minds. The principle of cognition on which it proceeds is utterly antagonistic to the Nominalism which denies all objectivity to genera and species: it is drawn from Realism alone, not the Scholastic Realism of the Middle Ages, but the Scientific Realism or Relationism which will be explained below. Nominalism teaches that things conform to cognition, not cognition to things; Scientific Realism teaches that cognition conforms to things, not things to cognition. It is futile to seek a reconciliation of these positions; the contradiction is absolute and insoluble. Modern philosophy counts nothing as "known" which is outside of the individual consciousness; modern science presents as "known" a vast mass of truths, of which only an insignificant fraction can be to-day comprised within the narrow limits of a single consciousness, and which in their totality can be contained only in the universal mind of man. Under the influence of the all-prevailing Nominalism of the present day, philosophy has, and must have, its beginning-point in the individual Ego; under the influence of its own unsuspected Realism, science begins with a Cosmos of which the individual Ego is merely a part. The one is exclusively and narrowly subjective, just so far as it is logically faithful to its own clearly proclaimed principle of cognition; the other is objective, in a sense so broad as to include the subjective within itself. In truth, so far was the old battle of Nominalism and Realism from being fought out by the end of the fifteenth century, that it is to-day the deep, underlying problem of

problems, on the right solution of which depends the
life of philosophy itself in the ages to come. But let
it not be forgotten that the old Realism of Scholasticism is by no means the new Realism of Science; the
former perished as rightfully before Nominalism as
Nominalism itself will perish before the latter.

That the scientific point of view is a thoroughly
objective one, and that the cosmical facts discovered
by science can by no means be made to vanish in the
universal solvent of Nominalistic subjectivism, easily
appears. One or two illustrations will suffice.

· Prof. Jevons, in the *Principles of Science* (3d ed.,
pp. 8, 9), thus speaks of the objective validity of
mathematical formulæ : —

" A mathematician certainly does treat of symbols,
but only as the instruments whereby to facilitate his
reasoning concerning quantities; and as the axioms
and rules of mathematical science must be verified in
concrete objects in order that the calculations founded
upon them may have any validity or utility, it follows
that the ultimate objects of mathematical science are
the things themselves. . . . Signs, thoughts, and ex-
terior objects may be regarded as parallel and analogous series of phenomena, and to treat any one of the
three series is equivalent to treating either of the
other series."

Prof. Tyndall, in his *Light and Electricity* (pp. 60,
61), thus illustrates the unhesitating and uncondi-
tional objectivity with which the science of physics
presents its truths, as facts of a veritably existent
and actually known Cosmos : —

" The justification of a theory consists in its exclusive competence to account for phenomena. On such
a basis the Wave Theory, or the Undulatory Theory

of Light, now rests, and every day's experience only makes its foundations more secure. . . . This substance is called the luminiferous ether. It fills space; it surrounds the atoms of bodies; it extends, without solution of continuity, through the humors of the eye. The molecules of luminous bodies are in a state of vibration. The vibrations are taken up by the ether, and transmitted through it in waves. These waves impinging on the retina excite the sensation."

Prof. Cooke, in his *New Chemistry*, illustrates the same point still more strikingly and emphatically, with reference to the atomic theory : —

"The new chemistry assumes as its fundamental postulate that the magnitudes we call molecules are realities ; but this is the only postulate. Grant the postulate, and you will find that all the rest follows as a necessary deduction. Deny it, and the 'New Chemistry' can have no meaning for you, and it is not worth your while to pursue the subject further. If, therefore, we would become imbued with the spirit of the new philosophy of chemistry, we must begin by believing in molecules ; and, if I have succeeded in setting forth in a clear light the fundamental truth that the molecules of chemistry are definite masses of matter, whose weight can be accurately determined, our time has been well spent."

Remembering that the weight of the hydrogen-atom is taken as the unit of molecular weight, or microcrith, and that, according to calculations based on the figures of Sir William Thomson, this atom weighs approximately, in decimals of a gramme, 0.000,000,000,000,000,000,000,109,312, or 109,312 octillionths of a gramme, one can easily perceive the impossibility of construing this utterly unimaginable quantity under any terms expressive of human consciousness. To consciousness it is equivalent to absolute zero; but the "New Chemistry" demands belief in it as an actual quantity in Nature, an objectively existent reality in a Cosmos not resolvable into consciousness by any Nominalistic legerdemain.

It would be superfluous to cite further passages in order to illustrate the thoroughly objective spirit, method, and results of modern science, as contrasted with those of modern philosophy. All scientific investigations are founded on a theory diametrically opposed to that of Kant: namely, that things can be known, though incompletely known, as they are in themselves, and that cognition must conform itself to them, not they to it. This is the philosophical translation of the principle of verification. The Nominalism that inculcates the contrary doctrine is an excrescence upon modern philosophy, a cancerous tumor feeding upon its life. Science has achieved all its marvellous triumphs by practically denying the fundamental principle laid down by Kant, and by practically proceeding upon its exact opposite; and it is a scandal to philosophy that she has not yet legitimated this practical procedure, overwhelmingly justified as it is by its incontrovertible results. The time has come for philosophy to reverse the Roscellino-Kantian revolution, and give to science a theory of knowledge which shall render the scientific method, not practically successful (for that it already is), but theoretically impregnable. The present article is the beginning of an attempt in that direction. A glance at the course of speculation in the past will render clearer the nature of the problem which philosophy has now to solve.

II.

The pre-Socratic philosophy of Greece was unquali-
fied Realism, of a *naïve* and primitive type. The
earlier Ionic philosophers, Thales, Anaximander, and
Anaximenes, sought only to generalize the phenomena
of the outer world, as products of a single original
cause or principle (ἀρχή) — water, undifferentiated
chaotic matter (τὸ ἄπειρον), air, — but they never
dreamed of doubting its objective existence. The
Pythagoreans sought the causal unity of the universe
in its most general relations, as number, proportion,
harmony, order, law, which they conceived as at once
the abstract and concrete directive force of nature;
their cosmology was no less objective than that of
their predecessors. The Eleatics, Xenophanes, Par-
ménides, Zeno of Elea, Melissus, maintained the
principle of objective Monism; their ἕν καὶ πᾶν was
illimitable and immutable Being, devoid of every
positive attribute save that of thought, while the
manifold appearances under which it presents itself
to man were only mere seeming and delusion. But
there was no element of subjectivism in their cosmology; they attributed to the Cosmos permanence
without change, unity without multiplicity, as its
constitutive objective principle. Heraclitus taught'
that the principle of all things was fire, as the type of
ceaseless and universal change (πάντα χωρεῖ), in opposition to the Eleatics; but his cosmology was none
the less objective because he discovered in it only
change without permanence, multiplicity without
unity. Empedocles sought to mediate between the
Eleatic and Heraclitean views by positing four changeless elements, air, earth, fire, and water, with two con-

stant forces, love and hate, and by conceding endless
change in the combinations and mutual relations of
these permanent factors of creation; but he was
wholly as realistic and objective as his predecessors.
The Atomists, Leucippus and Democritus, offered a
strictly mechanical explanation of Nature, attributing
independent objective reality to the atoms which
alone remained changeless in the midst of eternal
change. Anaxagoras in a certain sense summed up
all the preceding philosophies in his own, by means
of his theory of ὁμοιομέρειαι or *semina rerum*, while he
introduced a new principle in the assumption of an
immaterial νοῦς as the moving and guiding cause of
the universe; and he, too, was unreservedly objective
in his cosmology.

With the Sophists, however, appeared the first
symptoms of true subjectivism; and they may be
regarded as the forerunners of Nominalism, though
only in a feeble, crude, and undeveloped sense. The
Sophists had no system, no school, no determinate
principle save that of scepticism as to objective truth
and paradoxical acquiescence in all opinions as equally
true or equally false. Their movement was the de-
structive distillation of all fixed conviction in the
heats of logomachy and interminable word-quibbling.
They had nothing in common save a certain unity of
spirit and method — a spirit of universal scepticism,
and a method of adroit disputation by the employment of double meanings and ambiguous middle
terms. Sceptics in philosophy, anarchists in ethics,
their greatest historical merit is that of having polar-
ized and called into activity the noble intellect of
Socrates. They held no definite theory of subjectivism at all; but the manner in which they evacuated

general terms of all fixed meaning and all objective
validity challenged and arrested the attention of
Socrates, as the true secret of their plausibility and
bewildering success in debate. It was this fact that
fixed and determined the direction taken by this
mighty genius. The Sophists practically, though not
theoretically, anticipated the Nominalists in conceding only subjective validity to generic and specific
terms, which constitute the very alphabet of knowledge ; and Socrates, piercing to the ulterior consequences of this procedure in the dissolution of all
intellectual verity and all moral obligation, rose, like
a giant in his strength, to combat a great tendency of
his time which threatened to cause the fatty degener-
ation of Greek civilization, the melancholy decay of
Greek thought and life.

The astounding success of Socrates in this great
struggle is the most splendid monument to the power
of individual genius that the history of philosophy
can show. Alone and unaided, he checked and reversed the Nominalistic revolution already far advanced, annihilated the Sophists as a practical power
in philosophy, and determined the course of speculation for a millennium and a half in the direction of
Realism. No other victory such as this was ever won
in the annals of human thought ; and yet what historian of philosophy has perceived, much less celebrated
it ? It will never be appreciated until the dominant
Nominalism of modern philosophy has given place to
the dawning New Realism of modern science — a day
perhaps less distant than now appears. What gave
success to Socrates in this vast encounter was the fact
that he planted himself on an immovable rock, the
objective significance and validity of general terms,

2

as opposed to their purely subjective import and
value. Even Schwegler, blind as he is to the enor-
mous importance of the struggle between Nominalism
and Realism (to which in his *History of Philosophy*
he devotes less than one page !), says of Socrates that
"there begins with him the *philosophy of objective
thought*" (p. 38, Stirling's translation — the italics
are his). Aristotle explicitly declares in the *Metaphysics* (XII. 4) that " Socrates was engaged in forming systems in regard to the ethical or moral virtues,
and was the first to institute an investigation in
regard to the universal definition of these. . . . There
are two improvements in science which one might
justly ascribe to Socrates — I allude to his employment of inductive arguments and his definition of the
universal. . . . Socrates did not, it is true, constitute
universals a thing involving a separate subsistence,
nor did he regard the definitions as such ; the other
philosophers, however, invested them with a separate
subsistence." But Socrates did attribute universal
objective authority to the virtues he defined ; he
refuted the Sophistic construction of them as merely
subjective ; he repudiated the Sophistic notion that
nothing is good or bad by nature (φύσει), but only by
statute (νόμῳ), and vindicated the objectivity of
general terms *in some sense*, without reaching that
luminous doctrine of the objectivity of relations
which alone explains it clearly. That Socrates conceived of universals as objective realities, without
arriving at any definite conclusions as to the mode of
this reality, sufficiently appears from the subsequent
course of Plato and Aristotle, both of whom inherited
from Socrates the undefined objectivity of universals,
and each of whom proceeded to define it in his own

way. The point to be here specially noted is the fact that Socrates rolled back the advancing tide of Nominalism let loose by the Sophists, accomplished the feat by means of the definition of universals as objectively valid and real, and stamped the thought of fifteen hundred years with the impress of his own Realism.

The impending Nominalistic revolution having been thus definitely arrested by Socrates, — the great question of universals having been bequeathed by him to succeeding generations for a full and final solution, — the existence of an objective outer world was a common and undisputed premise among his followers. In particular, the assumption of the objective reality of genera and species, as necessarily involved in that of a cognizable outer world, and as constituting the objective ground of all general terms, became a common point of departure to Plato and Aristotle. But, while Plato erected on this assumption his theory of Ideas, Aristotle erected on it his opposing theory of Essences or Forms — to which reference will be more particularly made below. Both the Platonic and Aristotelian points of view were fundamentally and equally objective, and equally alien to the point occupied by modern philosophy since the triumph of Nominalism over Realism, when the tides of thought began to set irresistibly in the direction of subjectivism.

The Stoics betrayed to some extent the influence of the Sophists in their theory of universals. They discarded alike the Platonic theory of Ideas and the Aristotelian theory of Forms, and were apparently the first to proclaim distinctly the doctrine of subjective concepts, formed through abstraction. This doctrine, however, did not attain in their hands a full logical development into the theory of Nominalism; in fact, it did not at all prevent the Stoics from advancing to the construction of a positively objective cosmology and theology of their own; and, although with a serious logical inconsistency, they maintained on the whole an objective point of view.

The Epicureans, with their doctrine of the atoms and the truth of all perceptions of matter, may be considered quite free from the tendency to subjectivism, so far as the present discussion is concerned.

The Sceptics — the earlier with their " Ten Tropes," and the later with their " Five Tropes " — did not so much deny the existence of an outer world as the trustworthiness of human knowledge of it, and advanced no definite doctrine respecting universals. They occupied mainly negative and critical ground, and exerted no great influence in that controversy. Their arguments mostly rest on the assumption of Realism.

During the third great period of Greek philosophy, including the Græco-Judaic, the Neo-Pythagorean, and the Neo-Platonic schools, the predominant tendency was pre-eminently objective, since the mystical or theosophical contemplation of a Divine Transcendent Object by means of the " ecstatic intuition " is incompatible with an exclusive subjectivity. Theosophy, in fact, tends to reduce the subject to a state of pure passivity, and to absorb him completely in contemplation of the Object of worship.

In no period of Greek philosophy, therefore, did the Nominalistic tendency gain much force or headway after it had once been checked by Socrates. Its hour had not yet come.

Passing now to the Christian Era, it may be said that the Patristic period was devoted to the development of systematic or dogmatic theology, without interference from pagan philosophy after the closing of the School at Athens, in A.D. 529, by edict of the emperor Justinian. Since dogmatic theology, by the very nature of its conceptions, is unqualifiedly objective, the Patristic and in main the Scholastic periods are chiefly noticeable here as having carried the principle of objectivity to so abnormal and oppressive a degree of development as to cause speculation to rebound to the opposite extreme. The creation of a great body of doctrine held by the Catholic Church to be the absolute and unmixed truth of God, and the terrible intolerance with which the Church stamped out all dissent from this fixed standard of belief, inevitably tended to excite a reaction against it, in proportion to the mental activity of the age. Moreover, the Church had planted itself in philosophy upon the Realism of Plato and Aristotle; and it was equally inevitable that the reaction should be against this, no less than against the theology of the Church. There is no room for wonder, then, at the fact that the cause of Nominalism came to be identified with the cause of intellectual and religious freedom, and the triumph of the one with the triumph of the other. Consequently it is to the Scholastic period, and to the rise of the great controversy between Realism and Nominalism — the former representing Catholic orthodoxy and the latter heterodoxy, — that must be traced the beginning of the general subjective movement of modern philosophy, although this movement did not gain full headway till after the downfall of Scholasticism, when victorious Nominalism had time to develop unrestrained all the latent tendencies it involved. Tennemann has significantly and truly said that this momentous controversy was "never definitely settled." The reason is that both sides were right, yet neither wholly so; they did but bequeath to later times a problem they could not solve. Disguised as it is by new forms and new names, the immeasurably important issue between objectivism and subjectivism involved in that ancient controversy survives to-day. Nominalism, by virtue of the truth it contained and the freedom it represented, conquered Realism in philosophy, and culminated in the splendid genius of Kant; Realism, by virtue of the truth it too contained, conquered Nominalism in science, created an army of experimental investigators of Nature, and culminated in the establishment of the scientific method, which, though as yet purely practical and empirical, demands with increasing emphasis from philosophy a theory of knowledge that shall justify it in all eyes. Here is the explanation of the wide divergence, the virtual divorce and even antagonism, which is so patent a fact to all who look beneath the surface of things, between science and philosophy. All the intellectual interests of mankind must suffer greatly, until the breach is effectually healed; and the first step to the reconciliation so much to be desired must be a clear comprehension of the causes which have created the division. Hence the necessity of surveying the ancient battle-field of Scholasticism.

The proximate origin of the great mediæval dispute over the nature of universals seems to have been a passage at the commencement of Porphyry's *Introduction* to Aristotle's treatise on the *Categories*, known at the

time only through the Latin translation of Boëthius, in which these three problems were stated, but not elucidated, with respect to genera and species:—
"1. Whether they have a substantive existence, or reside merely in naked mental conceptions. 2. Whether, assuming them to have substantive existence, they are bodies or incorporeals. 3. Whether their substantive existence is in and along with the objects of sense, or apart and separable." Neglecting minor distinctions, refinements and subtilties, and without following the long and tedious course of the dispute, it will amply suffice for present purposes to state concisely the five leading positions maintained by different philosophers of the Scholastic period, as follows:—

1. EXTREME REALISM (*Universalia ante rem*) taught that universals were substances or things, existing independently of and separable from particulars or individuals. This was the essence of Plato's Theory of Ideas, and Plato was the father of Extreme Realism as held in the Scholastic period. Scotus Erigena, who died A.D. 880, was the first to revive this doctrine in the Schools, borrowing from the Pseudo-Dionysius Areopagita.

2. MODERATE REALISM (*Universalia in re*) also taught that universals were substances, but only as dependent upon and inseparable from individuals, in which each inhered; that is, each universal inhered in each of the particulars ranged under it. This was the theory of Aristotle, who held that the τόδε τι or individual thing was the First Essence, while universals were only Second Essences, real in a less complete sense than First Essences. He thus reversed the Platonic doctrine, which attributed the fullest reality to universals only, and a merely "participative" reality to

individuals. Until Scotus Erigena resuscitated the Platonic theory, Aristotle's was the received doctrine in the Schools; and the warfare was simply between those two forms of Realism prior to the advent of Roscellinus.

3. EXTREME NOMINALISM (*Universalia post rem*) taught that universals had no substantive or objective existence at all, but were merely empty names or words (*nomina, voces, flatus vocis*). Though probably not the absolute originator of this *sententia vocum*, as the doctrine came to be called, Roscellinus, Canon of Compiègne, was the first to give it currency and notoriety, and the Council of Soissons, under the influence of the Realist Anselm of Canterbury, his chief opponent, forced him in the year 1092 to recant the tritheistic interpretation of the Trinity, which he had consistently and courageously avowed. The theory of Extreme Nominalism was thus put under the ecclesiastical ban.

4. MODERATE NOMINALISM or CONCEPTUALISM (*Universalia post rem*) taught that universals have no substantive existence at all, but yet are more than mere names signifying nothing; and that they exist really, though only subjectively, as concepts in the mind, of which names are the vocal symbols. Abailard is claimed by some, but probably incorrectly, as the author of this modification of the Nominalistic view; William of Occam, who died in 1347, seems to have been the chief, if not the earliest, representative of it. The *Encyclopædia Britannica* (XVI. 284, 8th ed.) says : "The theory termed Conceptualism, or conceptual Nominalism, was really the one maintained by all succeeding Nominalists, and is the doctrine of ideas generally believed in at the present day."

5. Albertus Magnus (died 1280), Thomas Aquinas (died 1274), Duns Scotus (died 1308), and others, fused all these views into one, and taught that universals exist in a three-fold manner : *Universalia ante rem,* as thoughts in the mind of God; *Universalia in re,* as the essence (quiddity) of things, according to Aristotle; and *Universalia post rem,* as concepts in the sense of Moderate Nominalism. This is to-day the orthodox philosophy of the Catholic Church, as opposed to the prevailingly exclusive Conceptualism of the Protestant world.

Thus both Extreme Realism and Moderate Realism maintained the objective reality of genera and species; while both Extreme Nominalism and Moderate Nominalism maintained that genera and species possess no objective reality at all.

In contrast with all the views above presented, another and sixth view will now be stated, which, taken as a whole and with reference to the vitally important consequences it involves, is believed to be both novel and true.

6. RELATIONISM or SCIENTIFIC REALISM (of which *universalia inter res* may be adopted as an apt formula) teaches that universals, or genera and species, are, *first,* objective relations of resemblance among objectively existing things; *secondly,* subjective concepts of these relations, determined in the mind by the relations themselves; and, *thirdly,* names representative both of the relations and the concepts, and applicable alike to both. This is the view logically implied in all scientific classifications of natural objects, regarded as objects of real scientific knowledge. But, although empirically employed with dazzling success in the investigation of Nature, it does not

appear to have been ever theoretically generalized
or stated.

This view rests for its justification upon a broader
principle; namely, that of the *Objectivity of Relations*,
as opposed to the principle of the *Subjectivity of Relations*, which is the essence of the Nominalistic doctrine
of universals inculcated by modern philosophy. Kant
distinctly made "Relation" one of the four forms
of the logical judgment which determine the twelve
"categories of the understanding;" *i.e.*, the *à priori*
forms of thought, totally independent of "things-in-
themselves," and applicable to them only so far as
they are objects of a possible "experience," which,
however, reveals nothing of their real nature. This
doctrine that relations do not inhere at all in "things-
in-themselves," but are simply imposed upon them by
the mind in experience as the purely subjective form
of phenomena, is strictly deducible from the Nominalistic doctrine that general terms, by which relations are expressed, correspond to nothing objectively
real; and Kant's master-mind is nowhere more clearly
apparent than in the subtilty and profundity with
which he thus seized the prevalent but undeveloped
Nominalism of the modern period, and erected it
into the most imposing philosophical system of the
world. By this doctrine of the Subjectivity of Relations, Kant reduced the outer world to utterly unknown *Dinge-an-sich*, and paved the way for his still
more thorough-going disciple, Fichte, to deny their
very existence, and thereby to take a great stride in
conducting Nominalism to its only logical terminus,
Solipsism.

The principle of Relationism, however, rests on
these self-evident propositions: —

1. Relations are absolutely inseparable from their terms.

2. The relations of things are absolutely inseparable from the things themselves.

3. The relations of things must exist where the things themselves are, whether objectively in the Cosmos or subjectively in the mind.

4. If things exist objectively, their relations must exist objectively; but if their relations are merely subjective, the things themselves must be merely subjective.

5. There is no logical alternative between affirming the objectivity of relations in and with that of things, and denying the objectivity of things in and with that of relations.

For instance, a triangle consists of six elements, three sides and three angles. The sides are things; the angles are relations — relations of greater or less divergence between the sides. If the sides exist objectively, the angles must exist objectively also; but if the angles are merely subjective, so must the sides be also. To affirm that the sides are objective realities, even as incognizable things-in-themselves, while yet the angles, as relations, have only a subjective existence, is the *ne plus ultra* of logical absurdity. Yet Kantianism, Nominalism, and all Nominalistic philosophy (if they admit so much as the bare possibility of the existence of things-in-themselves) are driven irresistibly to that very conclusion.

In short, it is because modern philosophy rests exclusively on the basis of Nominalism, of which the only logical terminus is absolute Egoistic Idealism or Solipsism, and because modern science rests exclusively on the basis of Relationism, that we affirm

unqualifiedly an irreconcilable antagonism between
the two just so long as their respective bases remain
unchanged. It seems needless, but may be nevertheless advisable, to point out explicitly that Relationism
carefully shuns the great error of Scholastic Realism,
i.e., the hypostatization of universals as substances,
entities, or things; it teaches that genera and species
exist objectively, but only as relations, and that things
and relations constitute two great, distinct orders of
objective reality, inseparable in existence, yet distinguishable in thought.

The philosophic value of the principle of Relationism is strikingly illustrated in the ease with
which, applied as a key, it unlocks the secret and
lays bare the signification of the ancient and
still unfinished controversy between Realism and
Nominalism.

1. It shows that Extreme Realism was right in
upholding the objectivity of universals, but wrong
in classing them as independent and separable substances or things.

2. It shows that Moderate Realism was right in
upholding the objectivity of universals, but wrong
in making them inherent in individuals AS INDI-
VIDUALS (*in re*) rather than in individuals AS GROUPS
(*inter res*). Relations do not inhere in either of
the related terms taken singly, but do inhere in all
the terms taken collectively.

3. It shows that Extreme Nominalism was right
in denying the objectivity of universals as substances or things (the great error of its opponent),
and right in affirming the existence of universals
as names; but wrong in denying their objectivity
as relations and their subjectivity as concepts.

4. It shows that Moderate Nominalism or Conceptualism was right in denying the objectivity of universals as substances, and also right in affirming their subjectivity as concepts; but wrong in denying their objectivity as relations.

Thus every element of truth is gathered up, and every element of error is eliminated, by rejecting the four historic theories already recapitulated, together with the merely syncretistic fifth theory, and by substituting in their place the propounded sixth theory of Relationism. Its precision, lucidity, comprehensiveness, and adequacy to account for all the facts, will become so evident to any one patient enough to master it fully in all its bearings, as to warrant the indulgence of a hope that it may permanently solve the great problem declared by Tennemann to have never been "definitely settled."

III.

When Scholasticism fell, the theory of Relationism had occurred to no one. Each of the competing theories discerned the weakness of its rivals, yet could not discern its own, and was therefore unable to arrive at the real truth respecting universals. Consequently, as has just been pointed out, the truth was divided among them. Nominalism gradually won the ascendency among philosophers in the form of Conceptualism; while Relationism became, not indeed a received theory, since as a theory it did not yet exist, but yet the unformulated and empirical principle of the actual practice of scientific observers, experimenters, and investigators of nature. Philosophy divorced itself from a true objectivity, and surren-

dered itself to subjectivism in the form of Moderate
Nominalism; while science, ceasing to philosophize,
turned its back upon the barren metaphysics of the
schools, because they could yield no objective knowledge, and learned the sad lesson of contempt for
philosophy itself.

A period of transition followed the downfall of
Scholasticism, full of confusion and conflicting tendencies. Spasmodic resuscitation of various ancient
philosophies — Aristotelianism in a more accurately
known form, Platonism, Neo-Platonism, Stoicism, Epi-
cureanism, &c. — ensued; but these revived systems
did not materially contribute to the growth of the
subjective tendency, since, as has been shown, ancient
philosophy in the post-Socratic periods had been prevailingly objective in all its forms. The true origin
of the increasing subjectivism of philosophy, and
therefore the true secret of the increasing repugnance
of science for philosophy itself, lay in the triumph
of Nominalism over the relatively inferior Realism of
the Middle Ages, in its denial of all objective knowledge save of particulars *as isolated and unrelated,* and
in its claim of a strictly subjective genesis for universals as concepts or names alone. Philosophy in
this manner stripped the objective world of everything that was really intelligible — genera, species,
relations of all kinds; while science, bereft of all philosophical aid, took refuge in a rude sort of common
sense and fortified itself in a spirit of defiance to all
speculative thought. Bacon's popularity rested really
on no stronger foundation: he merely headed an un-
reasoning revolt against Nominalism, hardly knowing
what he did, yet practically rendering an immense
service by rallying the enterprising and curious spirits

of the time about the standard of "induction." He too joined in the wide-spread outcry against Aristotle and his followers, mistakenly believing that Aristotle was really responsible for the Nominalism of the age which he vaguely felt to be the chief obstacle to science. The results of this open feud between science and philosophy were disastrous to both in the end; for, while the latter tended steadily towards Idealism and Solipsism, the former as steadily tended towards Materialism. For the time being, however, the revolt of science against philosophy was most salutary.

While science adopted a purely empirical objective method, took Nature for granted, investigated things and their relations by observation and experiment on the hypothesis of their equal objectivity, and entered on a career of dazzling conquest, without troubling itself to invent any philosophical justification for a method so prolific of discoveries as to silence all criticism or cavil by the brilliancy of its achievements, philosophy had already entered upon a path which led indeed to the construction of numerous subjective systems of unsurpassed ability, yet to none that could endure. The history of philosophy has been for three centuries only a succession of gayly-colored pictures, each more startlingly beautiful than the last, yet each doomed to disappear at the next turn of the kaleidoscope. While science can proudly point to a vast store of verified and established truths, which it is a liberal education to have learned and the merest lunacy to impugn, philosophy has achieved nothing that is permanently established. The cause of this vast difference in result is a radical difference in method. Objectivism, albeit solely empirical, has created the glory of science; subjectivism, albeit elaborately and ostentatiously reasoned, has created the shame of philosophy. And philosophy can never redeem itself from this shame of utter barrenness until it repudiates subjectivism with Nominalism, its cause.

.The epoch of Scholasticism is regarded by some as closed by the death of Gabriel Biel, the "last Scholastic," in 1495, when Nominalism had acquired almost undisputed sway.

Now the essential method of Scholasticism had been, as Tennemann well expresses it, to "draw all knowledge from conceptions." So long as Realism flourished, and universals, as entities, were held to possess substantial objective existence, the analysis of concepts, independently of experience or verification, was held to yield real knowledge of their objective correlates — a mistake impossible to the New Realism or Relationism. But when Nominalism had destroyed the objectivity of universals, it had also destroyed the possibility of deriving objective knowledge from concepts. A dilemma thus arose: either objective knowledge is unattainable, or it must be attained otherwise than by the mere analysis of concepts as such. But how?

In this manner was developed a new and momentous problem, that of the Origin of Knowledge, which now displaced the old and still unsolved problem of the Nature of Universals — not at all fortuitously, but logically and inevitably as a direct result of the triumph of Nominalism. Nominalism had answered the old question after its own manner by resolving universals into merely subjective notions; and this answer, false as it was, was accepted as satisfactory.

But the acceptance of it involved some awkward consequences. If objective knowledge cannot be derived
from concepts, whence can it be derived? Or is
there no such thing as objective knowledge?

Science met these questions by boldly adopting the
principle of Objective Verification — a principle depending absolutely for its philosophical justification
on the theory of Relationism, but adopted by Bacon
and the inductionists in general as a purely empirical
method, in utter indifference to such justification.
From that time forward, scientific men have quietly
assumed the objectivity of relations, and steadily
pursued the path of discovery in total disregard of
the disputes of metaphysicians — not, however, without a serious loss to science itself, in the growth and
spread of the false belief that science can legitimately
deal only with physical investigations, and that the
scientific method has no applicability in the "higher
sciences."

But philosophy met the same questions by dividing
into two hostile camps. The sufficiency of the Nominalistic answer to the question of universals — that
they are exclusively of subjective origin — was taken
for granted by both parties; genera, species, relations
of all kinds, were unanimously conceded to possess
no objective validity whatever. Logically, this is the
total surrender of all objective knowledge; and in
the long run modern philosophy has come to accept
this result, as shown by the almost entire unanimity
of modern philosophers in the opinion that things-
in-themselves, or noumena, are utterly incognoscible.
But it is impossible to maintain this opinion in logical consistency, and on this point not a single logically
consistent philosopher can be pointed out; if he can

be found, he will prove to be an inexorably rigorous Solipsist, not afraid to deny the existence of all minds save his own, no less than that of the material world. It would be refreshing to meet with a subjectivist possessed of the courage of his opinion; but he would be the terror of all his brother-subjectivists, perhaps a candidate for premature interment.

The division that now arose and separated modern philosophy into two great contending parties did not concern the question whether knowledge originated in the object or in the subject, — for both parties agreed in the Nominalistic answer to this question, — but whether, in the subject mind itself, it originated in the senses or in the intellect. That was the great new question started at the recognized dawn of modern philosophy by Descartes and Locke; and both parties to the controversy, both the *à priori* and the *à posteriori* schools, were equally switched off upon the false track of Nominalism that conducts to Egoism or to nothing.

Descartes' theory of "innate ideas" encountered a vigorous rival in Locke's theory of experience as limited to the data of "sensation and reflection;" and thus the two armies took position for the long warfare that is resultless still. There is not the slightest occasion, for the purposes of this paper, to follow the course of this dispute, or to repeat the argumentation and counter-argumentation by which it has been maintained. The point of view here taken is that both these famous schools have logically immured themselves in the dungeon of subjectivism, and are utterly powerless to release themselves; that the one is just as incompetent as the other to explain the "origin of knowledge" about which they have been contending so long; and that, like Venus and Mars suspended in Vulcan's cage to provoke the "inextinguishable laughter" of the Odyssean gods, they do but enact a farce at which philosophy hangs her head. Travelling round the same circle of subjectivism in opposite directions, these two schools are fated to re-unite on the farther rim in one identical point — the stand-point of Absolute Egoistic Idealism. That is the only possible terminus of a subjectivism that, beginning with the definition of knowledge as only the mind's recognition of its own states, dares to obey the logic of its own fundamental principle; and what is the philosophy worth that contradicts itself ? No sequent thinker who begins with the Ego as sole starting-point will fail to end with the Ego as sole terminus, unless he stoops to unworthy tricks or evasions; and that is the suicide of philosophy.

The triumph of Nominalism did indeed force upon thought a new problem in the question of the "origin of knowledge;" but great is the delusion of the two schools which imagine the solution of that question to lie with one of themselves.

The *à priori* school started with Descartes' *Cogito ergo sum;* that is, with an original positing of the Ego as an *individual thinking being.* The *à posteriori* school started with Locke's "sensation;" that is, with an original positing of the Ego as an *individual feeling being.* That is essentially the only difference — the difference between beginning with individual thought or individual feeling as the prior element of individual consciousness, — both beginnings being equally and incontrovertibly egoistic. But this is a trivial difference indeed, compared with the abysmal difference between both these egoistic schools, on the one hand, and modern science, on the other ; for here the issue is a broad, deep, fundamental one — namely, whether the real "origin of knowledge" is in the Ego or in the Non-Ego, or in both. Knowledge itself, in the conception of both these Nominalistic schools, is confined to the series of changes that go on in consciousness; and all their mutual discussions are mere child's-play, compared with the discussions that await philosophy the moment she comes abreast of the time.

Science is to-day challenging emphatically the very foundation of both *à priori* and *à posteriori* philosophies; and the challenge is none the less menacing or deep-toned, because it has been hitherto uttered in deed rather than word. She denies, not by a theory as yet, but by the erection of a vast and towering edifice of verified objective knowledge, that genera and species are devoid of objective reality, or that general terms are destitute of objective correlates; she denies that Nominalism has rightly solved the problem of universals, when that solution would in an instant, if conceded, sweep away all that she has won from Nature by the sweat of her brow. Her very existence is the abundant vindication of Relationism, as the stable and solid foundation of real knowledge of an objective universe. As the case now stands, philosophy has two great schools, equally founded on a reasoned subjectivism which *denies the possibility of knowing*, in any degree, an objectively existent cosmos as it really is; while science rests immovably on the fact that she *actually knows* such a cosmos, and proves *by verification* the reality of that knowledge which philosophy loudly and emphatically denies. Science must be all a huge illusion, if philosophy is right; philosophy is a sick man's dream, if science is right.

One or the other must speedily effect a total change of base; and it is safe to predict that the change will not be made by science.

Three answers are given, therefore, to the question as to the Origin of Knowledge; two by Nominalism, with its two schools of modern philosophy, and one by Relationism, interpreting the silent method of science. They are substantially as follows: —

1. The *à priori* school teaches that knowledge has two ultimate origins, the experience of the senses and the constitution of the intellect — the senses contributing its *à posteriori* "matter" and the intellect contributing its *à priori* "form;" that the intellect is the source of certain universal and ante-experiential principles of knowledge which cannot be in any manner derived from the senses; that these principles or "forms" are themselves an object of pure *à priori* cognition, independently of experience; that experience consists solely of sense-phenomena, and sense-phenomena give no knowledge of their merely hypothetical noumenal causes, *i.e.*, of " things-in-themselves." In other words, things (if they exist — which is at least dubious) conform themselves to cognition; the subject knows only its own subjective modifications, arranged in a certain order according to *à priori* laws of knowledge which are only subjectively valid. This is Nominalistic Subjectivism of the *à priori* type.

2. The *à posteriori* school teaches that knowledge has only one ultimate origin, the experience of the senses; that the intellect is indeed the source of certain universal constitutive principles of knowledge, but that these were originally derived from the senses, having been slowly organized and con-

solidated, by the law of the "association of ideas," into hereditarily transmissible "forms" of experience; that there is no such thing as "pure *à priori* cognition," independent of experience; that experience consists solely of sense-phenomena, that the intellect itself has been slowly evolved out of it, and that sense-phenomena give no knowledge of their merely hypothetical noumenal causes. In other words, things-in-themselves (if they exist — which is equally dubious by this theory) conform themselves to cognition; the subject knows only its own subjective modifications, arranged in a certain order according to *à posteriori* laws of knowledge, which are only subjectively valid. This is Nominalistic Subjectivism of the *à posteriori* type.

Thus both of these dominant schools thoroughly agree in planting themselves upon the foundation of Moderate Nominalism or Conceptualism; they agree that universals, the genera and species by which alone sense-phenomena are reducible to intelligible order, are merely subjective concepts without objective correlates. They agree that things-in-themselves are unknown and unknowable, and that the subject knows its own conscious states alone. By both schools, consequently, the principle of Relationism is either unknown or ignored; relation itself is by both reduced to a merely subjective category, valid only as the subjective order imposed on subjective sense-phenomena, and utterly meaningless as applied to noumena; and noumena — intelligible objective realities, as presented by the various sciences — are totally incognoscible. But when the vitally pertinent question is put: " *Why* should the series of sense-phenomena, or sensations, or consciousness in general, be what

it is ? *Why* should the senses and understanding
conspire to give a coherent appearance of objective
knowledge, when no objective knowledge is possible?"
neither school has any reply to make. The only reply consistent with their common premises would be
Fichte's reply, that the apparent objects of knowledge
are given by the subject to itself, according to some
inscrutable law working subtly beneath consciousness
itself. This reply has at least the merit of consistency
with the ground-principles of subjectivism, and does
not flinch from landing philosophy in Solipsism undis-
guised. But few subjectivists possess sufficient hardi-
hood to make this consistent reply; they prefer to
"have their cake and eat it too."

3. The theory of Scientific Philosophy (by which
is meant simply the philosophy that founds itself
theoretically upon the practical basis of the scientific
method) teaches that knowledge is a dynamic correlation of object and subject, and has two ultimate
origins, the cosmos and the mind; that these origins
unite, inseparably yet distinguishably, in experience,
i.e., the perpetual action of the cosmos on the mind
plus the perpetual reaction of the mind on the cosmos
and on itself as affected by it; that experience, thus
understood, is the one proximate origin of knowledge;
that experience has both an objective and a subjective
side, and that these two sides are mutually dependent
and equally necessary; that the objective side of experience depends on the real existence of a known
universe, and its subjective side on the real existence
of a knowing mind; that experience includes all mutual
interaction of these, whether sensitive or cognitive,
and is utterly inexplicable even as subjective sensation, unless its sensitive and cognitive elements are

equally recognized; that this extended conception
of experience destroys the distinction of noumena
and phenomena, as merely verbal and not real; that
"things-in-themselves" are partly known and partly
unknown; that, just so far as things are known in
their relations, they are known both phenomenally
and noumenally, and that the possibility of experi-
mentally verifying at any time their discovered relations is the practical proof of a known noumenal
cosmos, meeting every demand of scientific certitude
and furnishing the true criterion and definition of
objective knowledge. In other words, science proceeds upon a principle diametrically opposite to that
of Nominalism, already explained under the name of
Relationism. It assumes that cognition conforms
itself to things, not things to cognition, — that
being determines human thought, not human thought
being, — that the subject knows not only its own
subjective modifications but also the objective things
and relations which these modifications reveal. Kant
did but "assume" the counter-principle; and if he
considered his assumption as at last "demonstrated"
by his system as a whole, science equally considers
its assumption as demonstrated by the actual existence of its verified and established truths as a body
of objective knowledge.

These three answers to the question as to the origin
of knowledge show how vast is the divergence between
modern philosophy and modern science. Philosophy
has never yet entirely shaken off the blighting influence of Scholasticism, even while fancying itself
wholly emancipated from it; for Nominalism, no less
than the old Realism, was the legitimate offspring
of Scholasticism. It was only one of the two great

answers, both one-sided and both wrong, which Scholasticism gave to the question of universals. Philosophy is still Scholastic to-day; it has never yet modernized itself in any true sense, and it never will do so until it sits modestly at the feet of science, imbues itself thoroughly with the spirit of the scientific method, and applies the principle of Relationism to the reconstitution of the moral sciences and the total reorganization of human knowledge. This, though a vast revolution for philosophy herself, will be simply giving in her adhesion to the revolution which science made long ago, and has rendered irreversible. But it will also be putting herself at the head of that revolution, and conducting it to conquests in regions of the highest truth of which science herself has never yet dreamed.

IV.

Aristotle taught, with truth, that the proper object of science is the universal rather than the particular or individual. Although it was his doctrine that individuals are First Essences, while species are Second Essences, and genera Third Essences, real only in a lower sense than the former, nevertheless it was also his doctrine that the universal inheres in each individual substance and constitutes its conceptual or intelligible essence (ἡ κατὰ τὸν λόγον οὐσία). The universal and the individual were inseparable, and must therefore be known together: yet the universal, being the essence of the individual, was itself the only proper and real object of scientific cognition.

Translating the Moderate Realism of Aristotle into the more accurate language of Relationism, and not

forgetting to correct its capital error of making the
universal inhere in each individual as an individual
(*in re*) rather than in all the individuals as a group
(*inter res*), the meaning of his doctrine is that science
is concerned with the general relations of things rather
than with the things themselves — with general laws
rather than with the peculiarities or accidents of
individual objects.

Modern science proceeds uniformly according to this
incontestable principle. Says Prof. Jevons: —

"There is no such process as that of inferring from
particulars to particulars. A careful analysis of the
conditions under which such an inference appears to
be made shows that the process is really a general one,
and what is inferred of a particular case might be
inferred of all similar cases. All reasoning is essentially general, and all science implies generalization.
In the very birth-time of philosophy this was held to
be so : 'Nulla scientia est de individuis, sed de solis
universalibus,' was the doctrine of Plato, delivered
by Porphyry. And Aristotle held a like opinion :
Οὐδεμία δὲ τέχνη σκοπεῖ τὸ καθ' ἕκαστον . . . τὸ δὲ καθ'
ἕκαστον ἄπειρον καὶ οὐκ ἐπιστητόν. 'No art treats of particular cases, for particulars are infinite and cannot be
known.' No one who holds the doctrine that reasoning may be from particulars to particulars can be supposed to have the most rudimentary notion of what
constitutes reasoning and science."

It is, in truth, impossible to study even a particular
case without generalizing; all knowledge consists in
the seizure of the relations of things, and every name
of a relation is of necessity a general term. Prof.
Jevons correctly quotes both Plato and Aristotle as
concurring in this fundamental principle, since both

of them occupied the standpoint of objectivism; and Prof. Jevons himself, as a scientific man, can occupy no other, although, as a thinker more or less infected with the subjectivism of modern philosophy, he has not succeeded in occupying it always or with entire consistency.

Now subjectivism reduces all science to the knowledge of *one individual*, the Ego, — which, as just shown, is no science at all. If its fundamental definition of knowledge means anything, or is faithfully adhered to, subjectivism teaches that the intelligent subject has no intelligence save of itself — has no warrant for believing in the existence of anything save itself — knows nothing but the inexplicable order of its own sensations and thoughts. It reduces all existence to an unrelated One, while of an unrelated One no science is possible. In a word, subjectivism, if logical, annihilates science at a blow.

There is no logical escape from this inference, drawn directly from the subjectivist definition of knowledge. Subjectivism cannot concede the knowledge of any existence except that of the subject itself; it cannot concede any knowledge of the subject, except that of its seriated conscious states; it cannot concede any knowledge of these conscious states as a series, but only as single and unrelated; and it thus lands us ultimately in the scepticism of Hume. For to generalize a series of thoughts as *thought*, or a series of sensations as *sensation*, is to use a general term, which, *ex hypothesi*, corresponds to no existent correlative in an objective sense; the general terms, thought, sensation, consciousness, on the principle of Nominalism, denote nothing real in the thoughts, sensations, or consciousnesses which are generalized, but express only an act of the subject as generalizing. Apply the very same principle to the knowledge of the subject itself which subjectivism applies to the knowledge of the outer world, — refuse that objective validity to general terms as applied to the world of consciousness which is refused to general terms as applied to the world outside of consciousness, — and it is shown irresistibly that subjectivism does not permit "knowledge" even of the subject's own "conscious states." "Consciousness" is a general term; "state" is a general term; every such term denotes a relation among certain related objects; and if this relation must be separated from the related objects when they are *outside of the subject*, why must it not be separated from the related objects when they are *within?* Subjectivism necessarily destroys itself by its own definition of knowledge; it cannot exist an instant except by denying the very principle it asserts; it escapes self-annihilation only on the hard and humiliating condition that it shall perpetually contradict itself. The sword with which it slays science pierces its own heart.

Nothing is more astonishing than the utter indifference of subjectivists to their own innumerable self-contradictions on these vital points — self-contradictions all the more amusing in view of their insistence that objectivism shall be rigorously and consistently reasoned. Let a few instances be here noticed.

Berkeley's idealism (a direct product of the Nominalistic revolution) is usually praised to the skies as unerringly logical and self-consistent. Yet the same reasoning which leads him to deny the existence of a material world ought to lead him to deny the existence of other human minds — of which there is no

proof except sight, hearing, and touch of the material
bodies by which these minds manifest themselves.
Berkeley's great paralogism on this point is pointed
out even by his own editor, Dr. Krauth (p. 400), as
follows : —

"Berkeley is a realistic idealist, holding that the
realistic inference is invalid as regards matter, but
conceding it as regards mind. He holds to real
substantial spirits, God and man. Hence, too, his
monism is only generic. He holds to· a monism of
genus, — to spirit alone ; but he concedes a dualism
of species, — infinite Spirit, the cause of ideas, and
finite spirits, the recipients of them. But this his
strength is also his weakness. Every moral advan-
tage of his Idealism over its successors is secured at
the expense of its development and of its logical
consistency."

Dr. Shadworth H. Hodgson, in his *Time and Space*
(Introduction, p. 5), says : —

"By the term consciousness, in this Essay, is always
meant consciousness as existing in an individual conscious being; and proofs drawn from such a consciousness can have no validity for other conscious
beings, unless they themselves recognize their truth
as descriptions applicable to the procedure and phenomena of their own consciousness. Doctrines, if
true, will ultimately be recognized as such by all
individuals whose consciousness is formed on the
same type, that is, by all human beings."·

Here is luminously presented the cardinal and
universal contradiction in all non-solipsistic forms
of subjectivism: (1) The assumption that the Ego
knows only the changes of its own consciousness ; and
(2) the assumption that the Ego knows other Egos to

exist that are "formed on the same type." One of
these assumptions necessarily destroys the other.

There are countless similar self-contradictions scat-
tered all through the writings of subjectivists, some
amusing by their *naïveté*, some ingenious in their
subtilty, some amazing by their evident unconsciousness, but all sufficiently humiliating and mortifying
to those who would fain see philosophy comport herself with the dignity of science rather than with the
agility of a circus-clown. One further illustration
will suffice.

Prof. Clifford, in his *Lectures and Essays* (II. 71),
takes the ground of the most uncompromising subjectivism at the outset, and then coolly proceeds to break
loose from it in the most violently illogical style, yet
apparently without the least suspicion of the exhibi-
tion he thereby makes of himself as a philosopher : —

"The objective order, *quâ* order, is treated by
physical science, which investigates the uniform relations of *objects* in time and space. Here the word
object (or *phenomenon*) is taken merely to mean a
group of my feelings, which persists as a group in
a certain manner; for I am at present considering
only the objective order of my feelings. The object,
then, is a set of changes in my consciousness, and
not anything out of it. . . . The inferences of physical science are all inferences of my real or possible
feelings; inferences of something actually or potentially in my consciousness, not of anything outside
of it."

Bald and unblushing as is the egoism of this
passage, it is entirely clear; and it is quite possible to build up on this basis an idealistic Solipsism
which shall at least tolerably cohere with itself. But

Prof. Clifford immediately proceeds to crucify his own subjectivism in this manner: —

"However remote the inference of physical science, the thing inferred is always a part of me, a possible set of changes in my consciousness bound up in the objective order with other known changes. But the inferred existence of your feelings, of objective groupings among them similar to those among my feelings, and of a subjective order in many respects analogous to my own, — these inferred existences are in the very act of inference *thrown out* of my consciousness, recognized as outside of it, as not being a part of me. I propose, accordingly, to call these inferred existences *ejects*, things thrown out of my consciousness, to distinguish them from *objects*, things presented in my consciousness, phenomena. . . . How this inference is justified, how consciousness can testify to the existence of anything outside of itself, I do not pretend to say: I need not untie a knot which the world has cut for me long ago. It may very well be that I am myself the only existence, but it is simply ridiculous to suppose that anybody else is. The position of absolute idealism may, therefore, be left out of count, although each individual may be unable to justify his dissent from it."

This airy distinction of "object" and "eject" does not in the least disguise the cardinal contradiction into which Prof. Clifford, in common with all subjectivists who shrink back from Solipsism, falls. Ejects, as he proceeds to define them, are simply "other men's minds;" but other men's minds are only known through their bodies, and their bodies are "objects" like trees or stones; while trees and stones are just as truly "ejects" from consciousness

as are other men's minds. In a word, ejects are
objects, and objects are ejects; there is absolutely
no distinction between them, on Prof. Clifford's own
showing; objects and ejects must be both objective
or both subjective. Yet Prof. Clifford arbitrarily (it
would almost seem wilfully) objectifies ejects and
subjectifies objects! He flatly refuses to "untie a
knot" which contains the whole point in dispute,
and which the "world" has "cut" just as effectively
for objects as for ejects; he coolly begs the whole
question, and repudiates the Solipsism from which his
own principles permit no rational escape.

These illustrations of the self-contradiction of subjectivism are *typical*, not *sporadic;* they show how
deep-seated is the disease under which modern philosophy is suffering. Whenever (if ever) subjectivism
shall dare to be rigorously logical, it will be the *reductio ad absurdum* of Nominalism, and compel philosophy
to adopt Relationism and the scientific method in general. All science is of the universal; all sequent subjectivism abolishes the universal, and leaves only the
individual, a solitary, unrelated, incomprehensible
Ego. It avails nothing to create a phantom-science
of the universal in a world of sensations alone; true
philosophy, no less than true science, demands an
explanation of that series of sensations which subjectivism can accept only as an unintelligible fact.
Diogenes commanded a certain respect so long as he
actually lived in his tub; but if, having fastened to
his forehead a placard, "I am Diogenes, and I live
in this tub," he had then tied the tub to his back,
lived in a house, slept in a bed, and behaved like
ordinary mortals, he would have been pelted with a
storm of pitiless gibes from the keen-witted Athenians.

And when philosophy, having tied the tub of subjectivism to its back, lives and lectures in a world of "ejects," and expounds to them a science of the objective relations they bear to each other and to an intelligible cosmos, human nature must have radically changed if philosophy fares any better.

It all comes to this ; either the truth of subjectivism or the truth of science is a pure illusion. The possibility of the one is the impossibility of the other.

The conclusion just stated finds abundant corroboration in contemporaneous thought. Subjectivism in philosophy has created a new type of scepticism in science. Urged as it were by a consciousness that it can only maintain its own truth by discrediting the truth of science, philosophy does not hesitate to undertake the task. Hence it has formulated a law of philosophical scepticism under the name of the "relativity of knowledge," founded upon a truism, but distorted into a falsity. Unable to shake the conviction of the reality of a known objective universe, and therefore unable to take the field in its only logical form of Solipsism, subjectivism nevertheless covertly saps the truth of science in a manner which hides its own fatal inconsistency. It declares that all knowledge is merely relative to human faculties, and it adroitly pushes this principle as if relativity were unreality. A quotation from Mr. Frederic Harrison's essay on "The Subjective Synthesis" will well illustrate the mode of its attack :—

"The truly relative conception of knowledge should make us habitually feel that our physical science, our laws and discoveries in Nature, are all imaginative creations — poems, in fact — which strictly correspond within the limited range of phenomena we have before

us, but which we never can know to be the real modes
of any external being. We have really no ground
whatever for believing that these our theories are the
ultimate and real scheme on which an external world
(if there be one) works, nor that the external world
objectively possesses that organized order which we
call science. For all that we know to the contrary,
man is the creator of the order and harmony of the
universe, for he has imagined it."

This subjectivistic scepticism, be it remembered,
has its root in the Nominalism which universally
prevails in philosophic circles, and which has profoundly affected those scientific men who, being more
than mere specialists, have felt their influence; and
it shows exactly where science must seek aid from a
renovated philosophy, if it is to escape suffocation by
the fire-damp of scepticism engendered by its own
operations. "If every genus is only a mere word,"
says a writer in the *Encyclopædia Britannica*, "it
follows that individuals are the only realities, and
that the senses are at bottom the only sources of
knowledge. And not only so, but on this theory no
absolute affirmation respecting truth is possible, for
such an affirmation involves of necessity a general
idea, which *ex hypothesi* is destitute of real validity.
Hence we have scepticism at the next remove." Mr.
Harrison is an illustration of the literal accuracy of
this statement. But the case is not bettered if the
genus is "only a mere" *concept*, instead of "only a
mere word;" for Extreme Nominalism and Conceptualism (the latter of which this writer accepts) are
equally sceptical in their implications, since they
equally disown the objectivity of relations. Only
the theory of Relationism fully meets the case.

The doctrine of the "relativity of knowledge,"
under cover of which subjectivism makes its attack
on the objective truth of science, undoubtedly rests
on a truism : namely, that knowledge is itself a relation between the knowing and the known, and that
nothing can be known except as it is known by the
knowing faculties. This, surely, is a very innocent
proposition. It simply means that man cannot know
everything; it does not at all mean that he does not
know what he knows. That human knowledge of the
cosmos is incomplete, partial, inadequate, could be con-
troverted only by a consistent subjectivist, to whom
the cosmos is simply the sum of his own sensations or
consciousness, which, again, exist only as they are
known. But the doctrine of the relativity of knowledge, properly construed, has a real validity and profound significance to the objectivist, since it states
the fact on which the total activity of science rests
— the fact that human knowledge is small, and can
be increased. There is nothing whatever in this doctrine to discourage science or impugn the solid character of its acquisitions. From the very nature of
the case, nothing but relative knowledge is possible.
Increase the number and scope of man's cognitive
faculties till his science becomes omniscience: his
knowledge will still be relative, being the relation of
knowing and known, and that unconditionally. In
fact, "non-relative knowledge" is a contradiction *in
adjecto.* As Prof. Ferrier puts it in his *Remains:*
" To know a thing *per se,* or *sine me,* is as impossible
and contradictory as it is to know two straight lines
enclosing space; because mind by its very law and
nature must know the thing *cum alio, i.e.,* along with
itself knowing it." The doctrine of the relativity of

knowledge, therefore, is a truism so far as it asserts the co-essentiality of subject and object to the relation of knowledge ; it is a falsity and absurdity so far as it asserts the non-knowableness of the object by the subject in that very relation of knowledge. And the blade of subjectivism is shivered in its very grasp by the adamantine shield of science.

Nevertheless it remains true that the progress of science is retarded and embarrassed by the prevalence of a philosophy which secretly undermines its results, controverts its fundamental postulate of the knowableness of the objective universe, and dooms it to an imperfect comprehension of the principles which alone justify its practical procedure. A philosophical vindication of those principles which should establish the scientific method, so resplendently successful in its empirical employment, upon an impregnable rational theory, could not fail in ten thousand ways to promote the advancement of knowledge, and dissipate that cloud which hangs over the deeper thought of our own age — the cloud of an intellectual consciousness at war with itself. Every attempt in this direction should be greeted with a hearty welcome.

Let us review the situation, and state the problem distinctly which philosophy has now to solve.

Subjectivism in philosophy takes its stand, consciously or unconsciously, on Nominalism. Its fundamental principle is the law, accepted by both the Transcendental and Associational schools, that things conform themselves to cognition, not cognition to things. The necessary corollary of this law is the *separability of phenomena and noumena,* phenomena having their existence solely as modifications of the individual consciousness, and noumena either having no existence at all or else existing solely as the unknown and unknowable causes of phenomena. Of these two alternatives, the former alone is logically consistent with the premises of subjectivism; for, since "cause" is a universal term to which Nominalism denies all objective validity or significance, it is a term patently inapplicable to anything beyond the sphere of subjective consciousness. Hence the final outcome of all thoroughgoing subjectivism is absolute egoistic Idealism or Solipsism — a mere cosmos of objectively causeless dreams.

Objectivism in science takes its stand, consciously or unconsciously, on Relationism. Its fundamental principle is the law of Objective Verification, — that cognition must conform itself to things, not things to cognition. The necessary corollary of this law is the *inseparability of noumena and phenomena*, phenomena being the "appearances" of noumena, and noumena being that which "appears" and is partially understood in phenomena; and they have their inseparable existence, not only in the mind, but also in the cosmos which the mind cognizes. The only utility in retaining the distinction at all is to mark the distinction between complete and incomplete knowledge — noumena being taken to denote things-in-themselves as they *exist* in all the complexity of their objective attributes and relations, and phenomena being taken to denote these same things-in-themselves so far only as they are *known* in their objective attributes and relations. The final outcome of scientific objectivism is a constantly growing knowledge of the real cosmos as it is, in which the human mind has its proper place and activity in entire harmony with cosmical laws.

This is the unequivocal issue between the two modes
of viewing the universe which are confusedly and half-
consciously struggling for supremacy in the modern
mind. Philosophy is prevailingly subjective, but not
wholly so; there are occasional symptoms of secret
restiveness among philosophers under the iron yoke
of Nominalism, such as the appeal of the Scotch
School to "Common Sense," the "Natural Realism"
of Hamilton, the "Reasoned Realism" of G. H. Lewes,
the "Transfigured Realism" of Mr. Herbert Spencer,
the "Inferential Realism" of Rev. J. E. Walter and
many others, the unmistakably objective tendencies
of the historian Ueberweg — who explicitly declares
that "the objective reality of relations can be affirmed
with at least as much reason as it can be disputed"
(*Hist. Phil.* I. 374), and that "the demonstrative
reasoning by which we go beyond the results of isolated experience, and arrive at a knowledge of the
necessary, is not effected independently of all experience through subjective forms of incomprehensible
origin, but only by the logical combination of experiences according to the inductive and deductive
methods on the basis of the order immanent in things
themselves" (*Ibid.* II. 162), — as well as of others
that might be named in this connection. But no one,
even among these uneasy insurgents against the established tyranny of Nominalism, seems to comprehend
exactly what the tyranny or who the tyrant is; no
one of them seems to have traced back the origin of
his oppression to the half-forgotten decision, arrived
at centuries ago by the now despised Schoolmen, as
to the nature of universals; and no one seems to comprehend precisely what will free him from fetters that
are invisible, yet strong as steel. Hence every one of

them continually falls into concessions which rivet
the fetters more closely about his limbs. The hostility secretly existing and working between the subjectivist and objectivist methods, even in one and the
same mind, is one of the curious and striking features
of contemporaneous thought, and will not fail to arrest
the attention of the future historians of philosophy.
Yet this antagonism between science and philosophy
is really unnatural and injurious in the last degree,
for they are the natural complements and allies of
each other. Science needs the intellectual order-
liness and systematic unity which philosophy alone
can create; philosophy needs the verified basis and
thoroughly objective spirit of science. Hence our age
presents no problem more profound in its nature, or
more wide-reaching in its bearings upon the intellectual interests of mankind, than this : —

*How to identify science and philosophy, by making
the foundation, method, and system of science philosophic, and the foundation, method, and system of
philosophy scientific.*

The theory of knowledge which is predominant
in both the Transcendental and Associational schools
of modern philosophy has been clearly set forth in
the preceding pages, traced to its source in the
wrong answer given by mediæval Nominalism to the
questions of universals, and shown to impart even
to so-called modern philosophy a thoroughly Scholastic character. The theory of knowledge which
underlies the practical procedure of modern science
has also been clearly set forth, although only so
far as its fundamental principle is concerned, under
the name of Scientific Realism or Relationism, — the
full development of which will involve the creation

of a new and comprehensive philosophical system.
The irreconcilable antagonism of these two theories,
the disastrous consequences of it both to philosophy and science, and the necessity of a profound
revolution in the method of philosophy in order to
bring it into harmony with the now thoroughly established scientific method, have likewise been shown,
together with the precise nature of the problem which
philosophy has now to solve, in order to modernize
itself in a true sense.

All that is here possible is simply to state the
problem and the general principle on which alone it
can be solved; a full solution of it is the great desid-
eratum of science and philosophy alike. For a full
solution of it will permanently heal the breach which
now disastrously divides them, and for the first time
render possible the harmonious co-operation and con-
centration of all the powers of the human mind for
the discovery, establishment, and application of cosmical truth. What has been here done is to show
that this greatest of modern problems is only, under
a new form, that ancient and never satisfactorily
answered question of Universals which, for hundreds
of years, absorbed the brightest intellects of Europe,
— to submit to the bright intellects of our own time,
together with the old half-answers to that problem
historically known as the theories of Nominalism and
Realism, a third, new, and full answer in the theory
of Relationism, — and to inquire whether this theory
will not suffice to bring about the greatly needed
identification of Science and Philosophy.

PART I.

THE PHILOSOPHY OF SCIENCE.

PART I.

THE PHILOSOPHY OF SCIENCE.

———•———

CHAPTER I.

THE PRESUPPOSITIONS OF THE SCIENTIFIC METHOD.

§ 1. MODERN science consists of a mass of *Propositions*[1] respecting the facts, laws, order, and general constitution of the universe. It is a product of the aggregate intellectual activity of the human race, and could no more have been produced by an individual than could the language in which its propositions are expressed. These propositions incorporate the results of universal human experience and reason, from which all elements of personal eccentricity, ignorance, or error have been

---

[1] "The answer to every question which it is possible to frame must be contained in a Proposition, or Assertion. Whatever can be an object of belief, or even of disbelief, must, when put into words, assume the form of a proposition. All truth and all error lie in propositions. What, by a convenient misapplication of an abstract term, we call a Truth, means simply a True Proposition; and errors are false propositions. . . . The objects of all Belief and all Inquiry express themselves in propositions." (John Stuart Mill, *System of Logic*, I. 18–19, London, 1872.)

gradually eliminated in the course of ages; they are
the winnowed grain of knowledge, from which the
chaff of individual mistake has been blown away by
the wind of universal criticism, and comprise the
total harvest of truth thus far garnered by man in
the study of Nature. All propositions respecting
the universe, whether in its physical or psychical
aspect, which at last command the unanimous assent
of all experts in the subjects to which they relate,
take rank as *Established Scientific Truths* — not necessarily as infallible truths, but as truths which stand
unchallenged until the progress of discovery compels a revision, correction, and re-establishment of
them as still larger truths. Infallible truths are
not for fallible man, and modern science is no more
infallible than ancient science; yet science is man's
nearest approximation to the absolute truth itself,
since it rests on no individual or dubious authority,
but on the highest possible authority which the
nature of the case permits: namely, the universal
experience and reason of mankind, voiced in the
unanimous consensus of the competent.

§ 2. Now all the established truths which are
formulated in the multifarious propositions of science have been won by use of the *Scientific Method.*
This method consists essentially in three distinct
steps : (1) observation and experiment, (2) hypothesis,
(3) verification by fresh observation and experiment.
Observation and experiment consist in the discovery, by actual perception, of things and relations

objectively existent in the universe, and constitute
that original experience of the universe in which
all human knowledge begins. Hypothesis, or the
rational interpretation of the results of observation
and experiment, is the ideal or subjective anticipation of further possible experience of the universe;
in its legitimate scientific use, it is the work of
reason and imagination combined, elaborating the
data of experience both inductively and deductively,
and inferring from already known relations other
relations which may objectively exist in the universe, and which, therefore, may be experientially
discovered there. Verification is the conversion of
sagacious hypothesis into theory and scientific law,
by means of fresh and corroborative experience;
what is verified is hypothesis, proved to have been
well-founded as inference, whenever the set of relations inferred is discovered by actual experience to
be identical with the corresponding set of relations
in the objective universe;[1] and the perception or
discovery of this identity, which is the essence of
all verification, proves that the constitution of the
universe and the constitution of the human mind
are fundamentally one. *Experience*, therefore, is the
beginning and the end of the scientific method,
mediated by reason and imagination; and experience itself is the actual meeting, the dynamic correlation, the incessant action and reaction, of the

---

[1] This is substantially Spinoza's test of truth: " Idea vera debet
cum suo ideato convenire." (*Ethica*, I. Ax. 6.)

human mind and its cosmical environment. The
scientific method, therefore, is a living organic process, the true and only organon for the discovery
of truth ; and the proof of its validity is the rapid
progress of actual discovery in the experiential study
of the universe.

§ 3. Now the scientific method logically implies a
very definite *Philosophy*, which it does not stop to
prove, but takes for granted and presupposes at
every step. In the course of many generations of
individual investigators, it has produced, as I have
said, a vast mass of propositions or established scientific truths, dealing directly with the facts and laws
of the universe itself, — not at all with men's ideas
of the universe, as ideas. For instance, astronomy
and physics make known various real relations
among real masses moving in real space, in absolute
independence of man, his existence, and his consciousness ; physics and chemistry make known
various real relations among real molecules and
atoms, likewise moving in real space; biology
makes known various real relations among real
living organisms; physiological psychology (which
sometimes mistakes itself for philosophy, but is in
fact one of many special sciences) makes known
various real relations between the physical system
and psychical activities of the individual organism ;
sociology, political economy, jurisprudence, ethics,
make known various real relations among human
individuals co-existing in a state of society. In

other words, the same scientific method, variously
applied in the various sciences, makes known (if
the word knowledge denotes anything but an impossible dream) a vast mass of *objectively real relations among objectively real things* — things and
relations which, although undeniably known by
consciousness alone, do not, for all that, depend
upon it in the least for their existence, inasmuch
as many of them are known to have existed mill-
ions of ages before human consciousness began.

An "objective," or "objectively real," or "objectively existent" relation must be understood simply
as a relation which subsists in the real universe
itself, and is not a mere conception of the human
mind. A relation may be known to exist objectively, whenever the proposition asserting it is
proved by experience to be *true.* For instance,
"the earth and the moon revolve about their common centre of gravity" expresses an objectively real
relation, because the scientific method has discovered
that such is the fact, independently of man, — that
the proposition is true. But the relation must not
be misconceived as a "thing," nor the affirmation of
the objectivity of the relation as an affirmation that
the relation is an entity apart from the things it
relates. *The known objectivity of a relation is simply
the known objective truth of the proposition which states
it.* But the relation itself was objectively real before
the proposition which states it was conceived; it determined the proposition, not the proposition it.

§ 4. It is evident, therefore, that the validity of the scientific method, and the objective truth of the results won by its use, depend unconditionally on the truth of the following philosophical presuppositions, which are never formally mentioned in any particular scientific investigation, or formally stated as part of any particular science, simply and solely because they are the common ground on which all science must stand, if it is to stand at all, and because they constitute the universal condition of the possibility of experience itself : —

PRESUPPOSITION I. An external universe exists *per se,* — that is, in complete independence of human consciousness so far as its existence is concerned; and man is merely a part of it, and a very subordinate part at that.

PRESUPPOSITION II. The universe *per se* is not only knowable, but known — known in part, though not in whole.

PRESUPPOSITION III. The "what is known" of the universe *per se* is the innumerable relations of things formulated in the propositions of which science consists; consequently, these relations objectively exist in the universe *per se,* as that in it which is knowable and known.

I repeat: the validity of the scientific method, the validity of the results won by its use, and the validity of these philosophical presuppositions, all stand or fall together ; for the presuppositions are nothing but a general explicit statement of what lies logically implicit in each of the numberless particular truths which constitute the body of science itself. It is not at any one's option to accept these particular truths, and at the same time reject the general statement which merely sums them up in brief. The actual existence of a universe independent of human consciousness, its actual intelligibility, and the actual existence in it of relations in which its intelligibility consists, — these, I maintain, constitute fundamental principles of a *Scientific Ontology*, presupposed at every step by the scientific method. Taken together and systematically developed, these principles will found a philosophy of science, embracing not only a radically new theory of knowledge, but also a radically new theory of being. The rapid disintegration of old philosophies, the widespread and growing confusion of religious ideas, and the universal mental restlessness which characterizes our age, are but the birth-throes of this new philosophy of science.

§ 5. It would be a very shallow criticism which should charge me here with returning to the old and unsatisfactory realism of the Scotch school, known as the "philosophy of common sense." Prof. Huxley, it is true, has described science as merely the extension and enlargement of "common sense," and he is not wrong in conceiving them as both realistic; but, if he had the Scotch school in mind, he disregarded the profound difference of the two with respect to the sources of their realism.

The Scotch school derived the conviction of the existence of an external world, not from scientific experience, but from a fundamental principle or "natural belief" originally implanted by God in the constitution of the human mind, and thus assigned to it a strictly *à priori* or subjective origin.[1] But the philosophy of science will derive it, not from any *à priori* constitution of the human mind, but from experience alone, corrected by reason, recast and elaborated by the scientific imagination, and verified by fresh experience, and will thus assign to it a strictly *à posteriori* or objective origin. Furthermore, the Scotch school held, not only that the things which we perceive exist, but also that they exist as we perceive them;[2] whereas the philosophy of science will hold that the crudities of sense-perception and the confused inferences of uninstructed

[1] "All the arguments urged by Berkeley and Hume against the existence of a material world are grounded upon this principle, that we do not perceive external objects themselves, but certain images or ideas in our own minds. But this is no dictate of common sense, but directly contrary to the sense of all who have not been taught it by philosophy." (Reid, *Intellectual Powers of Man,* Essay VI. chap. V.) "In the order of nature, belief always precedes knowledge. . . . Even the primary facts of intelligence, — the facts which precede, as they afford the conditions of, all knowledge, — would not be original, were they revealed to us under any other form than that of natural or necessary beliefs." (Sir W. Hamilton, *Lectures on Metaphysics,* p. 32, Amer. Ed.) "The doctrine which has been called *The Philosophy of Common Sense* is the doctrine which founds all our knowledge on belief." (Id. *Lectures on Logic,* p. 383.)

[2] "Another first principle is, That those things do really exist which we distinctly perceive by our senses, and are what we perceive them to be." (Reid, *l. c.*)

"common sense" are to be corrected by scientific
discovery, and will therefore present, as the veritable
outward fact, the subtile and often recondite relations which her formulated laws express. Lastly,
the Scotch school taught the mediæval doctrine of
Conceptualism or Nominalism,[1] which logically implies that of the merely subjective reality of relations ; whereas the philosophy of science will teach
the great principle of Relationism, which posits the
objective reality of relations as the cosmical corre-
late of universal concepts in the human mind — an
innovation sufficient of itself to revolutionize and
modernize the falsely so called "modern philosophy."
These, not to mention other important differences,
are quite enough to signalize the vast divergence between the philosophies of science and of "common
sense," and to show that scientific realism is of a
type wholly distinct from that of the Scotch school.

§ 6. Still more shallow, however, would be the
criticism that scientific realism is a mere groundless
assumption, an unreflective and untutored begging
of the question, a *naïve* taking for granted by "common thinking" of the whole point at issue : namely,
whether or not an external universe can be known
as independent for its existence upon human con-

---

[1] "The Doctrine of Nominalism has, among others, been em-
braced by Hobbes, Berkeley, Hume, Principal Campbell, and Mr.
Stewart ; while Conceptualism has found favor with Locke, Reid,
and Brown. . . . This opinion [Nominalism] . . . appears to me
not only true, but self-evident." (Sir W. Hamilton, *Lect. on Met.*,
p. 477.)

sciousness. On the contrary, scientific realism has
an inexpugnable rational foundation in the triumphantly successful use of the scientific method by
the separate sciences, and points out that this incontrovertible success has settled the question experi-
mentally, decisively, and forever; it grounds itself
avowedly on the truth of the discoveries which the
scientific method has made; it declares that the
truth of these discoveries, once admitted, demon-
strates that experience cannot be the product of
consciousness alone, but must be the product of consciousness and an external universe endlessly acting
and reacting upon each other — cannot be the sole
activity of the subject, but must be the co-activity
of the subject and the object in dynamic correlation;
and it declares that this interpretation of experience
must be unreservedly conceded, or else the validity
of the scientific method itself must be unreservedly,
boldly, and frankly denied.

The sharp issue is this: either an external world
independent of human consciousness is known to
exist, or else all human science is false. By no
logical subterfuge can this issue be escaped. If the
discoveries made by science are real or true discoveries, if the relations they reveal in the non-human
universe are real or true relations, then scientific
realism is no assumption, no begging of the question,
no taking for granted of the point at issue, but the
most absolutely *proved truth* which the intellect of
man has ever wrested from the mystery in which

he dwells. The claim of science to be real knowledge of a real and intelligible universe is the voice of the collective experience and reason of mankind; it is a claim so solidly grounded that the hardiest sceptic durst not call in question the particular truths of which that knowledge is the sum.

It is only when these particular truths are generalized as I have generalized them, — only when the generalization is put into the form of a definite philosophical principle of Scientific Ontology, — that the sceptic's voice is heard. But, if he would successfully challenge scientific realism as a philosophical first principle, he must first overthrow all the particular truths of which scientific realism is a mere restatement in general terms. Scientific realism is no more an assumption than is science itself; the two are one and the same. The ground here taken is that *the Successful Use of the Scientific Method is the Verification and Demonstration of Scientific Realism;* that scientific realism can be overthrown only by overthrowing the scientific method itself; and that it is time for speculative philosophy to recognize this position, to appreciate its tremendous strength, and to adopt it as its own foundation and point of departure. Until it shall do so, speculative philosophy will never become the creator of any deep or world-wide human conviction, never mould the faith of mankind, never command the religious allegiance of the many, but must remain what it is to-day — the closet-amusement and intellectual luxury of the few. So long as it persists in denying that experience is actual knowledge of a universe independent of human consciousness, — so long as it persists in seeking a knowledge of Being which shall be deeper or higher than experience can give, — just so long will mankind at large consider philosophy itself as an ingenious boy in the backwoods inventing a machine for perpetual motion, when all the civilized world knows that a machine for perpetual motion is impossible.

§ 7. " But," it will be asked, " do you seriously mean to defend the exploded doctrine that the universe is known as a Thing-in-itself, a *Ding-an-sich*, a Noumenon ? "

That is exactly what I mean. But I deny that the doctrine is exploded, and I also deny that it has ever yet been set forth in its true light. The realism of science is assuredly no invention of mine; and it can no more be exploded without exploding the whole fabric of science, than the foundation could be blown from beneath the Washington Monument without bringing the whole majestic column in ruins to the ground. For the last two or three centuries, the most fashionable philosophy has played the part of a Japanese juggler or acrobat, and performed logical feats requiring no small agility and dexterity, yet not conducing in any marked degree to the advancement of civilization. Beginning with Descartes's famous " I think, therefore I am," — that is, with the certainty of individual human consciousness as the one first fact and starting-point in all speculation, — and assuming, as regulative principle of procedure, that nothing can be certainly known except the contents of individual human consciousness, modern philosophy would, if it reasoned well, arrive at the conclusion that nothing can be either *known*, or *inferred*, or *conceived*, as existent outside of individual human consciousness. With such a point of departure and such a rule of procedure, the only logical conclusion is absolute solipsism, or the sole existence of the individual thinker; every form of inferential realism relies on a logically worthless inference (§ 67). But modern idealism tries in a thousand ways, ingenious as they are futile, to escape from the unavoidably solipsistic outcome of its own principles, to withdraw all attention from this its great intellectual sin against the first laws of logic, and to arrive at some mode of living amicably with the external world which it can neither suppress nor master: all of which is commendably amiable, but not quite satisfactory as a substitute for clear thinking.

§ 8. Now the root of modern idealism, whether in its transcendental or experiential form, is the theory of *Phenomenism* — the theory that nothing can be known except " phenomena," and that all phenomena depend for their existence on individual human consciousness alone. It is this theory of phenomenism, the life-principle of modern philosophy, which most formidably opposes the theory

of *Noumenism* (scientific realism or scientific ontology), the life-principle of modern science. This profound and fundamental issue between PHENOMENISM
AND NOUMENISM lies at the bottom of all other issues
of modern thought; it is the "previous question" in
all philosophical controversies; it is the imperfectly
seen, yet uneasily and vaguely felt turning-point,
or strategical centre, in the movement and self-
marshalling of all warring tendencies in the dis-
tinctively modern mind; it is the pitched battle-field
in a struggle which must end in a vast intellectual
revolution, wrought by the influence of modern science
upon so-called modern philosophy, by which philosophy will become truly *modernized* — taught, that is,
to exchange its old, worn-out, and merely traditional
Scholastic Method of sterile subjectivism for the new
Scientific Method so prolific of objective discoveries.
For Phenomenism is the historical product of the
Kantian "*Apriorismus;*" the Kantian "*Apriori-
mus*" is the historical product of mediæval Nominalism; and mediæval Nominalism is the historical
product, by a violent and extravagant reaction explicable as historical polarization, of the earlier
mediæval Realism, which the Catholic Church had
borrowed from Plato and Aristotle, and had rendered
intolerable in the Renaissance by abusing it to the
service of oppressive and unintelligible dogmas.[1]
This indisputable genealogy of phenomenism shows
that the issue between it and noumenism is, in

[1] See the Introduction.

truth, the everlasting issue between the past and the present, and that all the interests of modern intellectual progress are involved in its right decision. Consequently, it is necessary to devote considerable attention to it, although it will be impossible here to do more than touch on a few salient points of so vast a subject.

CHAPTER II.

THE THEORY OF PHENOMENISM.

§ 9. STRIPPED of unessential particulars, the most
advanced and fully developed form of phenomenism
may be tersely stated in these five main positions: —

1. The universe is only a phenomenon, and not a
noumenon or thing-in-itself.

2. This phenomenon-universe, like every minor
phenomenon, is only a mental conception or representation, deriving its whole existence from the
representing consciousness alone, and determined by
and depending upon absolutely nothing which is
external to that consciousness.

3. For philosophy, the sphere of Being is strictly
identical with the sphere of the phenomenon-universe,
and therefore with the sphere of human representation; no inference either to a noumenal subject or
to a noumenal object is philosophically permissible.
All the categories, even those of Reality, Existence,
and Being itself, are mere forms of relation within
the actual content of human representation, and
have neither validity nor application beyond it. The
sole legitimate aim of philosophy, limiting its scope

both as Theory of Knowledge and Theory of Being, is to investigate these immanent relations of representations as such, and rigorously to exclude all hypotheses as to possible realities not actually contained within them.

4. Since all the categories by which representations are internally determined, including the category of Relation, are themselves determined *à priori* by (and hence deducible from) the nature of the human understanding,[1] all possible relations are merely immanent determinations of human representations, schematized by the pure understanding and the transcendental imagination acting in concert. In other words, no relations are possible in any noumenal world which may be external to the representations. Hence, even if a noumenal world exists, it must possess in itself a non-relational or chaotic constitution, and therefore remain forever unintelligible *per se.*

5. The existence of a noumenon-universe, however, even if an abstract possibility, is an utterly inconceivable, groundless, and useless assumption. The noumenon is a mere hypostasis of the abstract unity of the "thing," which abstract unity is nothing

---

[1] ——"*weil der Verstand des Menschen von Natur so organisirt wird, dass,*" *u. s. w.* (Krug, *Encyklopädisch-philosophisches Lexikon,* I. 730.) This "natural organization of the human understanding" is to phenomenism an ultimate and inexplicable fact. In this fundamental point, phenomenism imitates the "*naïve* realism" which it professes to despise; for it rests at last, no less than the Scotch school, on the assumption of an ultimately inscrutable constitution of the knowing faculty.

but the *à priori* form of representation in general;
by hypostasis, this mere *à priori* form of thought
is illegitimately converted into a self-subsistent entity
or "thing-in-itself." Consequently, there is and can
be no perceptive understanding or intellectual intuition (*intellectuelle Anschauung*) by which this nonentity may be cognized.

§ 10. In short, phenomenism is the theory which
teaches that the universe is a phenomenon without
a noumenon, existing in the act of the individual
consciousness which represents it, and while it represents it, but otherwise having no existence which
can be either known, inferred, or conceived ; and,
consequently, that science is valid only in the realm
of actual experience — valid, that is, only as explaining the order and connection of actually existent
representations, whose true explanation must be
sought only in themselves, and not in a self-existent
universe. In other words, all the relations formulated in the propositions of science are absolutely
created by the mind which formulates them, and
exist only in that mind; they do not exist in any
universe independent of it, but have their whole existence in the human representations of which they
themselves are merely immanent determinations.[1]

The rational foundation of this whole theory, then,
lies in the principle that *relations have no objective*

[1] "Materialism . . . builds its theories upon the axiom of the
intelligibility of the world, and overlooks that this axiom is at
bottom only the principle of order in phenomena." (Lange, *History of Materialism*, II. 166, Boston, 1880.)

reality whatever, but exist solely and exclusively as the creative work of the human understanding. This exclusive *Subjectivity of Relations* is the genetic and essential principle of phenomenism, although not distinctly laid down as such by phenomenists, and evidently not discerned by them to be the fundamental logical ground of phenomenism itself, for the reason that it has been inherited by all schools of modern philosophy from mediæval Nominalism, and hence has never been subjected hitherto to a closely critical examination.[1] It constitutes, for all that, the whole pith and substance of phenomenism and its chief future significance in the history of philosophy; for it is the germinal presupposition from which all the other principles of phenomenism have been logically derived, and without which they would have no inner coherence or even intelligible meaning.

§ 11. Taken in the advanced form which has been presented above, the theory of phenomenism is based substantially, though with various modifications and improvements, on the Kantian philosophy; and it

---

[1] Even M. Fr. Paulhan, who writes an article on "La Réalité des Rapports" in *La Critique Philosophique* for April 30, 1885, has to destroy his own argument by taking his stand on phenomenism: "Nous nous plaçons ici sur le terrain du phénoménisme qui voit dans les faits, quels qu'ils puissent être, non pas l'ombre changeante et fuyante d'une substance inconnaissable, mais une réalité vraie, la seule réalité dont on puisse, en somme, s'occuper." It is manifest enough that M. Paulhan is defending only the *phenomenal* reality of relations in the representation, not their *noumenal* reality in the thing-in-itself.

meets us everywhere in the philosophical literature
of the day. Prof. Windelband says of it: "This
thought, that outside of representation there is
nothing with which science has to deal, is Kant's
gift of the gods to man; although to common thinking, to which nothing is more familiar than the distinction of representation and thing-in-itself, it must
appear to be what Jacobi, the champion of common
thinking, called it — Nihilism." And again: "This
Immanent Method of the theory of knowledge is
now justly considered to be Kant's supreme achievement." Riehl goes so far as to declare that the
Kantian philosophy essentially consists in this "immanent method" of discarding both noumenal subject and noumenal object as mere metaphysical
dreams, and refusing to consider aught beyond the
bare representation itself. Fortlage, on the other
hand, formulates this method as an attempt "to
resolve all cognitions into the process of cognizing,"
and characterizes it as "completed scepticism." [1] Just
as the Scientific Method rests on the presupposition
of the Objectivity of Relations, so the Immanent
Method rests on the presupposition of the Subjectivity of Relations; both presuppositions are assumed
without proof, and constitute the rational ground
of their respective methods, the pivotal principles of
Noumenism and Phenomenism as rival theories of

---

[1] See the valuable article by Prof. W. Windelband, of Zürich,
"*Ueber die verschiedenen Phasen der Kantischen Lehre vom Ding-
an-sich,*" in the *Vierteljahrsschrift für wissenschaftliche Philosophie,*
I. 224–266, Leipzig, 1877.

knowledge. What, then, shall be said of the theory of Phenomenism? Is it true?

§ 12. I consider the theory of phenomenism false, root and branch, — false in relation to the opposite theory of noumenism, which is proved true by the existence of science as actual and indisputable knowledge of a noumenal universe, and false in itself, because it contradicts itself in a most astounding way. Omitting here all other criticisms, and reserving these for another occasion, I rest my case for the present on these two objections, either of which, if substantiated, is overwhelmingly decisive.

§ 13. The first objection to phenomenism is that science is actual knowledge of a noumenal universe, and therefore refutes by its bare existence the phenomenism which denies the possibility of such knowledge, — on the sound principle of the old logical maxim: "*Ab esse ad posse valet, a posse ad esse non valet, consequentia.*"

§ 14. To break the force of this argument, phenomenism, of course, maintains that science is, and claims to be, nothing but knowledge of phenomena alone, — that it neither has, nor professes to have, any knowledge of noumena. It denies that "the discovery of new relations between phenomena within the sphere of consciousness" can "either prove or disprove the existence of that noumenal something which was the object of the keen Irish Bishop's brilliant polemic." It strenuously contends that Nature is nothing more than a "system of sense-ideas: " that is, a merely *subjective synthesis of real sensations* mutually related and reduced to order in representation by means of the schematism of the pure understanding, and not at all an *objective synthesis of real relations* in a universe independent for its existence on human consciousness. It asserts that "investigation of the laws of Nature proceeds upon a basis of observation and experiment, and observation and experiment have to do with the immediate object of knowledge " (*i.e.*, as evidently here intended, not the objectively existent thing, but the purely subjective mental representation of the thing, the *Vorstellung*), "and in no case with the 'substratum' or 'thing-in-itself.'" It affirms that "the only difference in the views of Nature taken by the ordinary scientific realist and the consistent idealist is, that the one regards objects as actually existing between the intervals of his perception, while the other attributes to them a merely potential existence " (*i.e.*, regards them as *actually non-existent*, the perception absolutely creating them and the cessation of perception absolutely annihilating them as actual existences, — which is, of course, the only possible meaning of the Berkeleian principle that the *esse* of objects is *percipi*).[1]

§ 15. Now the conception of science here presented, if it were not so common in phenomenistic

---

[1] The quotations in this paragraph are all taken from an ingenious article by Prof. G. S. Fullerton, entitled "The Argument from Experience against Idealism," in the *Journal of Speculative Philosophy*, October, 1884.

literature, and if it were not unfolded with such evident gravity, seriousness, and *naïveté*, would be aptly characterized as mere caricature, travesty, or broad burlesque. Every one of the propositions which formulate the results of the scientific method, and constitute in their totality the body of science, is, if valid at all, valid of "things-in-themselves," — that is, states relations among objective realities which have indeed been discovered by human perception, yet no more depend upon human perception for their existence than the coach in the fable depended on the fly for its motion. That, and that only, is what every scientific man *means* by his statements, and he would be indignant, if told to his face he did *not* mean it.[1] By means of consciousness, science discovers permanent relations among permanent things which depend on consciousness for nothing whatever, except for the discovery itself. Phenomenism may deny the discovery, if it will, but not distort it; it has no right to pervert facts and misrepresent science by pre-

[1] The "order of Nature" is never understood by strictly scientific men in the sense of the "mere order of my representations," which is the interpretation put upon it by phenomenism. Prof. Virchow, in Schliemann's *New Ilios*, refers to his own " *Gewohnheit der kältesten Objectivität.*" Prof. W. B. Taylor, in his masterly essay on "Kinetic Theories of Gravitation," published in the *Smithsonian Report* for 1876, says: "Our beliefs should always be based upon, and conform to, the observed order of Nature." Prof. L. E. Hicks, who fills the chair of geology in Denison University, says in his *Critique of Design Arguments*, p. 17: "The external order existed before the science which is based upon it." Volumes could be filled with precisely similar statements.

tending that the discovery relates merely to subjective human representations, when it relates in truth to an objectively real and self-existent universe.

§ 16. It is solemn trifling or elegant pleasantry of this sort which has degraded philosophy from its once proud rank of *scientia scientiarum*, and threatens to degrade it still further to that of *ignorantia scientiarum.* The friendship which phenomenism professes for science is a false and treacherous friendship; for phenomenism is the modernized form of the ancient Greek scepticism, and has merely given to the crude Pyrrhonic formula of the "unintelligibility of all things" (ἀκαταληψία, πάντα ἐστὶν ἀκατάληπτά, *nihil sciri potest, ne illud ipsum quidem*) a more subtile and refined form in the modern doctrine of the "unintelligibility of things-in-themselves" (*Unerkennbarkeit der Dingean-sich*). To both the ancient and modern scepticisms science makes one silent reply: she points to her undeniable discoveries and the method by which they have been won, as the unanswerable proof that knowledge of the noumenal universe is attainable by *experience.* Certain it is that phenomenism, the thoroughly systematized form which scepticism has assumed in modern times, lays the axe at the very root of all scientific knowledge of the universe, by astutely and covertly seeking to transmute it into a purely ideal product of the human mind, devoid of all truth or applicability beyond the human mind itself. But its blows will

continue to fall without effect, until it shall first
have attacked and destroyed the scientific method.
Unwilling to attempt openly, however, so formidable
a task, phenomenism prefers to assume the guise of
friendship, to concede ostensibly the validity of the
scientific method and its results, and then to under-
mine it secretly by interpreting these results as
" the discovery of new relations between phenomena
within the sphere of consciousness." In other words,
since such relations must depend absolutely for their
existence upon the continuance of the consciousness
in which they are discovered, and must therefore
cease to exist the moment they cease to be perceived,
phenomenism covertly denies, notwithstanding her
professions of friendship, that the scientific method
can effect any discovery of any fact that does not
begin and end with human consciousness itself.
Consequently, when science (as she does) formulates
countless relations as *objectively real in the universe
per se*, phenomenism, not venturing to contradict,
misinterprets and misrepresents them as only *subjectively real in the human mind.* Despite all dis-
guises, phenomenism thus shows itself to be the
secret and irreconcilable foe of science, and appears
as what Fortlage calls it, " completed scepticism."
In short, if phenomenism is true, science is false;
if science is true, phenomenism is false; and every
attempt to show the contrary misrepresents one,
or the other, or both. The first objection, therefore,
that phenomenism is refuted by the bare existence

of science, is substantiated for all who are convinced
that science is true; but its full strength will hardly
be felt before the theory of noumenism is positively
developed.

§ 17. The second objection to the theory of phenomenism is that it suicidally contradicts itself,
inasmuch as it claims to get rid of noumena altogether, and ends by giving us nothing else.

§ 18. In the first place, it maintains that "the
phenomenon-universe is only a mental representation, deriving its whole existence from the representing consciousness." Now a "mental representation"
is nothing but the *act of representing,* just as a thought
is nothing but the act of thinking, or as a mental image is nothing but the act of imagining; its
existence consists in the actual continuity of the
act, and ceases when the act ceases. Moreover, the
"representing consciousness," likewise, according
to phenomenism, which rejects the supposition of
a noumenal subject behind the consciousness, is
nothing but the *act of representing;* for nothing else
remains when the noumenal subject is suppressed.
Consequently, the statement with which we began,
if we now substitute in it these strictly equivalent
expressions, will read as follows: "The phenomenon-
universe is only an *act of representing,* deriving its
whole existence from the *act of representing.*" Consequently, the phenomenon-universe, thus reduced
to a mere act of representing, derives its whole
existence from itself — is therefore absolutely self-

subsistent and depends on nothing beyond itself —
is therefore a self-existent or self-caused reality, or
causa sui. Thus phenomenism, pretending to give
us a phenomenon-universe, has given us in fact a
universe which is pure noumenon, and nothing else ;
for, if a *causa sui* is not a pure noumenon, nay, a
very *noumenon noumenorum*, what is it ? Phenomenism, therefore, strange to say, ends by giving us a
noumenon-universe after all !

§ 19. In the second place, the Greek sceptic Karneades, founder of the third Academy, knew how to
analyze the representation ($\dot{\eta}$ $\phi\alpha\nu\tau\alpha\sigma\dot{\iota}\alpha$) without denying the reality either of the representing consciousness
($\dot{o}$ $\phi\alpha\nu\tau\alpha\sigma\iota o\acute{u}\mu\epsilon\nu o\varsigma$) or of the represented thing ($\tau\dot{o}$
$\phi\alpha\nu\tau\alpha\sigma\tau\acute{o}\nu$).[1] Phenomenism, however, conceding real
existence to the representation, denies it both to the
representing consciousness and to the represented
thing.[2] Hence the pure representation, since it
really exists, yet can exist neither in a noumenal
subject nor in a noumenal object, must exist really
in itself — in other words, must be, and be known
to be, a self-existent entity dependent on nothing

[1] " Um die Unmöglichkeit eines Kriteriums und der darauf sich
stützenden Ueberzeugung darzuthun, analysirt er die Vorstellung
und findet, dass dieselbe ein Verhältniss habe, sowol zu dem Gegen-
stande, durch den, als zu dem Subjecte, in dem sie entsteht." (Erd-
mann, *Grundriss der Geschichte der Philosophie*, L 164.)

[2] " Die Beziehung unserer Vorstellungen auf ein vorstellendes
Subject und auf ein vorzustellendes Object sind in dem reinen
Thatbestand des Vorstellungsinhaltes nicht enthalten, sondern
bereits Deutungsversuche zur Erklärung der Vorstellungen, die
eine durch die Categorie der Causalität, die andere durch diejenige
der Substantialität vermittelt." (W. Windelband, l. c., p. 259.)

outside of itself. But that is a noumenon, in the very sense in which phenomenism most vigorously denies its reality. Once more, therefore, phenomenism, promising us a representation which shall be pure phenomenon, ends by giving us a noumenon-representation after all!

§ 20. In the third place, phenomenism is severe on all attempts to convert abstractions into entities by that delusive process of thought called hypostasis. "The tenets of the old metaphysic," says Windelband, "consisted in the hypostasis of the *à priori* forms of thought (*Hypostasirung der Denkformen*); the assumption of things-in-themselves in general is the hypostasis of the ground-form of all representations." And he declares: "The hypostasis of the thought-forms is the essence of all dogmatism." The warning is salutary; but phenomenism immediately proceeds to despise and disregard it. For the retort is cogent and unanswerable that, if the hypostasis of the *thought-forms* is inadmissible in the old metaphysic, the self-evident hypostasis of the *thought-acts*, or *thought-functions*, is no less inadmissible in phenomenism itself. The representation cannot possibly be conceived as anything else than a mere act or functioning of the mind, a mere act of representing; and, by abstraction, to elevate this mere act or function into a self-subsistent phenomenon-in-itself is to hypostatize it, beyond the possibility of cavil or reply. The "*Hypostasirung der Denkformen*" is at least no worse than the *Hyposta-*

sirung der Denkacten; the phenomenon-in-itself is
at least as bad as the thing-in-itself, — in fact, it
becomes a thing-in-itself, since there is no distinction between phenomenon and noumenon, if the
possibility of separating them is once conceded.
Self-existent representations, or phenomena-in-them-
selves, are strictly indistinguishable from noumena,
or things-in-themselves. Not only, therefore, does
phenomenism, having promised to give us phenomena
alone, end by giving us noumena alone, but also it
caps the climax of self-contradiction by creating its
noumena through the selfsame process which, in the
old metaphysic, it gravely reprehends and repro-
bates — the process of *hypostasis!*

§ 21. Thus, turn which way it may, phenomenism
proves itself utterly unable to escape from the noumena it abhors, and powerless to hold fast by
"phenomena alone;" *for "phenomena alone" in-
stantly become noumena.* In vain it struggles to
evade the necessity of confessing that man knows
the self-existent: the bare fact that anything exists
at all is demonstration of the fact that something
exists of itself, and the one fact is no less necessarily known than the other. The essential and
avowed purpose of phenomenism, namely, to conceive the universe as only a phenomenon, is, therefore, quixotic, impossible, and self-contradictory to
the very verge of absurdity. It cannot be characterized with a more thoroughly scientific accuracy than
by a passage in that charming story-book, *Alice in*

Wonderland, — designed, it is true, for children, yet
not without occasional instruction for philosophers.
Alice has repeatedly encountered the famous and ever-
grinning " Cheshire Cat," and at last exclaims : —

" ' I wish you would n't keep appearing and vanishing so suddenly ; you make me quite giddy.'

" ' All right,' said the Cat ; and this time it van-
ished quite slowly, beginning with the end of the
tail, and ending with the grin, which remained some
time after the rest of it had gone.

" ' Well ! I 've often seen a cat without a grin,'
thought Alice ; ' but *a grin without a cat !* It 's the
most curious thing I ever saw in all my life.' "

When philosophy becomes fairyland, in which
neither the laws of nature nor the laws of reason
hold good, the attempt of phenomenism to conceive
the universe as *a phenomenon without a noumenon*
may succeed, but not before ; for it is an attempt
to conceive " a grin without a cat." Being satisfied,
therefore, that phenomenism is the most inconsistent
and unphilosophical theory to be met in the whole
history of philosophy, I turn now to the opposite
theory of noumenism, or scientific realism.

CHAPTER III.

THE THEORY OF NOUMENISM.

§ 22. KANT occasionally opposes the phenomenon (*die Erscheinung*) to the non-phenomenal (*das Nicht-Erscheinende*), but far more frequently to the noumenon or thing-in-itself (*das Ding-an-sich*). Now the first is a true, the second a false opposition ; and the reason why he failed to see that these two oppositions were not one and the same lies deep in the ground-plan of his system, — nay, in the far older nominalistic theory of universals, of which his system is simply the historical and logical culmination. This point must be at least briefly explained, for it concerns our subject vitally.

§ 23. The general purpose of the *Critique of Pure Reason* is to investigate the origin and laws of pure *à priori* knowledge, the possibility and reality of which Kant far too hastily assumed, inasmuch as all the instances he gives of it, if keenly scrutinized, betray at once the presence of strictly empirical elements. Under the influence of the traditional and still prevalent Nominalism, which reduces all general terms to mere subjective concepts and by implication denies the possibility of objective relations as their cosmical correlates, — and more particularly under the influence of the nominalistic Hume, whose incomplete subjectifying of the causal relation stimulated Kant's profounder genius to develop this partial subjectivism into a universal and systematic " *Apriorismus,*" — Kant found himself logically compelled to consider the category of relation itself as of purely subjective validity, and to see in it merely one of the four *à priori* forms of the logical judgment (Quantity, Quality, Relation, Modality) which determine the twelve "categories" or "pure concepts of the understanding." Into these categories or *à priori* forms of thought, as if into moulds, the formless matter of sensuous intuition is run, and thereby enabled to take the form of definite representations. All relations, as such, were thus conceived by Kant to be exclusively subjective in origin and nature, and to be impressed, so to speak, on the data of sensation as an exclusively subjective element in all cognition of objects of experience. In this manner the far-reaching principle of the *Subjectivity of Relations,* derived from the old nominalistic theory of universals and simply reduced by Kant to a scientific form, became incorporated as a vitally essential part in the Kantian system; and then, for the first time, was the foundation laid for a thoroughly systematic theory of phenomenism.

§ 24. Now relations as such are the specific and only direct objects of the intellect or understanding. Nothing else can be properly said to be understood;

nothing else can even be affirmed, because every proposition without exception is simply the statement of some determinate relation between its subject and its predicate. Consequently, all relations having been resolved by Kant into a purely subjective *addendum* to the objects of experience presented to the senses, the external world became straightway absolutely stripped of everything which is intelligible; things in themselves, being left utterly unrelated either to each other or to the human understanding, lapsed into the condition of virtual non-existence; nothing remained possible but to view the universe in itself as an utterly inscrutable and unintelligible blank, if indeed it existed at all — which unintelligible existence Kant, indeed, affirmed, but which his successors have either gravely doubted or boldly denied.

Now these facts perfectly explain how the word noumenon,[1] which originally, in Greek philosophy, signified "that which is intelligible," came to mean in Kantian and post-Kantian use the exact opposite: namely, "that which is unintelligible." This total inversion in the meaning of one of the most important words in the philosophical vocabulary is certainly a most extraordinary, significant, and instructive fact; and I venture to assert that no satisfactory expla-

---

[1] The Greek *voέω*, even in Homer, signified to *perceive with the mind*, as well as *with the eyes*. In Plato, τὰ *voούμενα* were *the objects of intellectual perception*, and hence, in general, " the intelligible; " although the derivative *voητός* more literally corresponded to the Latin *intelligibilis*.

nation of it can be given except the revolution of
thought by which, through the rise of nominalism,
the principle of the objectivity of relations was sup-
planted by that of the subjectivity of relations.   To
Socrates, Plato, Aristotle, and the schools derived
from them, relations were objective realities, either
separable or inseparable from objective individual
things ; they were in no sense impressed on objects
à priori by the understanding in the act of cognition ;
they belonged to the things in themselves, and made
the things intelligible.   This is the essential purport
both of the Platonic Theory of Ideas and of the
Aristotelian Theory of Essential Forms, whence
arose the distinction of the κόσμος νοητός and the
κόσμος αἰσθητός, the *mundus intelligibilis* and the
mundus sensibilis — one and the same world in itself,
as differently related to the understanding and the
sensibility, yet equally within the compass of both.
It is no less the essential purport even of the Greek
" Skepsis ; " for the fundamental difference between
the ancient and the modern scepticisms, unnoticed
even in the best histories of philosophy, yet easily
detected behind their statements, lies in the fact
that ancient scepticism rested on the assumption of
the objectivity of relations, while modern scepticism,
or phenomenism, rests on that of the subjectivity of
relations.   To show this in detail would require
more space than can here be spared for the purpose ;
yet a few facts may be cited which sufficiently and
unmistakably indicate it.

§ 25. Pyrrhon, the founder of the (improperly so called) sceptical " school," developed the general philosophical doubt occasioned by the mutual conflict of the various dogmatic systems of his time into what may be termed a negative dogmatism, whose chief tenet was the total incomprehensibility or unintelligibility of things (ἀκαταληψία). This tenet was avowedly based on the observed conflict of human opinions (διὰ τὴν ἀντιλογίαν, ἐκ τῆς διαφωνίας); and this observed conflict of human opinions can evidently be construed in only one way, — namely, as an *actually existent relation of antagonism*, objective to and independent of the observer, yet actually perceived and known by him as a ground of inference. Pyrrhon, therefore, as is self-evident, denied neither the objective reality of things, nor the objective reality of their relations, nor the subjective reality of some mode of discovering at least the particular objective relations on which he based his general conclusion : on the contrary, he manifestly *assumed* all this, without noticing that it upset the conclusion itself. What he denied was the possibility of discovering what the real relations of things are, on account of the absence of any trustworthy criterion of truth; and what he affirmed was that nothing can be known with certainty, because to every affirmation respecting things as they are in themselves its negation can be opposed with equal plausibility or strength (παντὶ λόγῳ λόγος ἴσος ἀντίκειται). This last proposition Sextos Empeirikos, with the addition of "as it appears to me" (ὡς ἐμοὶ φαίνεται) in order to avoid even negative dogmatism, declared to be the ground-principle of scepticism (ἀρχὴ τῆς σκεπτικῆς). It was, therefore, only on the *naïvely* conceded reality of actual and perceptible relations in the intelligible world, as objectively existent and really discoverable, although curiously enough claimed to be undiscoverable, that the Pyrrhonists inculcated abstinence from all assertion (ἀφασία) and suspense of judgment (ἐποχή) respecting the constitution of things as they are.

The Academics Arkesilaos and Karneades substantially agreed with Pyrrhon, but, in order to escape an absolute deadlock in the world of action, allowed probability (πιθανότης) as a practical guide in common life. Ainesidemos brought the "Skepsis" to its highest pitch of perfection by conceiving it not as denial, or even as mere doubt, but rather as investigation. The true sceptic does not permit himself to maintain, like the Academics, that there is no certainty, but only probability; that would be a dogma; he affirms not, denies not, doubts not, but investigates; the essential thing is to maintain nothing at all, and to permit to oneself the use of no expressions more dogmatic than "perhaps," "I do not decide," "it is possible," "it may be or may not be," and so forth. This settled hostility to that fixedness of conviction which is the inevitable result of all positive experience and scientific verification is, perhaps, the chief point of union between the

ancient " Skepsis" and modern phenomenism : it is
the most marked characteristic of both, and reveals
itself in phenomenism as that diseased habit of
mind which abhors nothing so heartily as fixed conclusions, stigmatizes them under all circumstances
as " dogmatic," and insists on treating even verified
scientific truth itself as a "mere hypothesis." In
order to give philosophic form to this tendency,
Ainesidemos drew up the famous ten "Tropes"
(τρόποι τῆς σκέψεως), or universal grounds for the
sceptical suspense of judgment, which were afterwards reduced to five by Agrippa. The most noticeable and paradoxical fact about them is that
every one of them involves a distinct and unequivocal recognition of the objectivity of relations : every
one of them is based on the *observed differences of*
things — disagreements in the constitutions of different animals or men, in the testimony of the senses
in general, in human institutions, customs, laws,
superstitions, or opinions, in the various conditions
and circumstances.of human life itself, and so forth.
Nay, the eighth (ὁ ἀπὸ τοῦ πρός τι), which in fact
covers the ground of the entire ten, explicitly alleges
the constant changes in the relations of things to
each other and to us (relations, therefore, which
must be both real and perceived) as a reason why
the permanent constitution of the things themselves
cannot be certainly known.   And Sextos declares in
terms that not only phenomena (φαινόμενα), *but also*
noumena (νοούμενα), are legitimate objects of scep-

tical investigation: the result of which investigation
being to find equal strength (*ἰσοσθένεια*) in opposite
conclusions as to both, the "Skepsis" conducts to
the desired suspense of judgment and consequent
peace of mind (*ἀταραξία*).

Nothing can be plainer, therefore, than the fact
that the Greek scepticism itself, — much more, then,
the other schools of Greek philosophy, — were all
founded upon the principle, assumed rather than
criticised and proved, of the objectivity of relations
and the intelligibility of noumena no less than of
phenomena; and that this principle of objectivism
or noumenism is the profoundest distinction between
Greek and modern philosophy, inasmuch as the
latter is almost universally based on the principle
of subjectivism or phenomenism. Alike to transcendental idealism, experiential idealism, and all
other forms of nominalistic philosophy in general,
relations have become mere subjective realities, inherent in the representations and absolutely dis-
severed from the world in itself, — which, like a
decapitated trunk, is now so far gone in decay as
to be indistinguishable from absolute nonentity.
While, however, modern philosophy has well-nigh
unanimously followed in Kant's footsteps, aban-
doned the old Greek foundation of the objectivity of
relations, and adopted the mediæval foundation of
scholastic nominalism or the subjectivity of relations,
modern science still stubbornly occupies the old
Greek ground of realism, and by her amazing, ever-

multiplying discoveries has already rendered it an absolutely impregnable fortress for the philosophy of the future.

§ 26. We are at last, therefore, in a position to understand how it happened that Kant, confessedly the greatest genius in philosophy since Aristotle, came to confound the true opposition between the . phenomenon and the non-phenomenal, on the one hand, with the totally false opposition between the phenomenon and the noumenon, on the other hand. In both the Greek and the German philosophies, the phenomenon is the Apparent, to which the Non-apparent is a true opposite; in the Greek philosophy, however, the noumenon is the Objectively Related and Intelligible, while in the German philosophy it has become, as I have just explained, the Objectively Unrelated and Unintelligible.

Consequently, in the Greek philosophy, there is no fundamental opposition between the phenomenon and the noumenon, since the Apparent and the Intelligible are quite compatible predicates of Being-in-itself; in fact, they are indispensable and inseparable predicates of it, inasmuch as only the Apparent can be intelligible and only the Intelligible can be apparent, — inasmuch, furthermore, as there is no contradiction, but perfect compatibility, between Being and Appearance or between Being and Thought. But, in the German philosophy, the noumenon having become identified with the Objectively Unrelated and Unintelligible, or "thing-in-itself," the phenome-

7

non became naturally and inevitably identified with
the merely Subjectively Related and Intelligible, or
"representation;" in other words, phenomena became wholly detached from the world of Intelligible
Being and wholly transferred to the world of Ideal
Thought. Nothing could be further from the truth,
. as modern science interprets it; but that, nevertheless, is the history of German idealism in a
nutshell.

In this manner an unavoidable opposition, — false
in itself, but logically drawn from the premises
latent in Nominalism, the mediæval and scholastic
philosophy grounded on the assumption of the subjectivity of relations, — has grown up and become
established in Germany between Being and Appearance, thing-in-itself and representation, noumenon
and phenomenon. Kant's second opposition between
the phenomenon (*Erscheinung*) and the noumenon
(*Ding-an-sich*) was, therefore, logical enough in his
own system and quite legitimate in his own use of
words — interchangeable, therefore, with his first
opposition between the phenomenon (*Erscheinung*)
and the non-phenomenal (*Nicht-Erscheinende*). None
the less unfortunate, however, have been the consequences of the grave error originated by his creation of this false opposition between the noumenon
and the phenomenon; for it has deepened the chasm
between modern philosophy and modern science, and
prevented the incalculable good which would have
resulted from their cordial co-operation. For, in

both the Greek and the scientific conceptions of the universe, there is no opposition whatever between the noumenon and the phenomenon; on the contrary, they are absolutely inseparable predicates of Being-in-itself, or the universe as both self-existent and intelligible. And philosophy itself can never recover its ancient influence and position as the supreme intellectual power in civilization and culture, until it has thoroughly revolutionized and modernized itself by adopting unreservedly the noumenism of modern science.

§ 27. While phenomenism, therefore, cleaves to the German conception, and views the universe as phenomenal only,—that is, as a purely subjective representation without any noumenal object, — noumenism cleaves to the old Greek conception, and views the universe as both phenomenal and noumenal. Here is brought out with perfect distinctness and clearness the fundamental difference between phenomenism, or German subjectivism, and noumenism, or ancient Greek and modern scientific objectivism. The former assumes, utterly without warrant in reason or experience, the *actual separability* of the phenomenon and noumenon, resolves the phenomenal universe into the merely subjective representation (*Vorstellung*), and denies all objective reality to the noumenal universe (*Ding-an-sich*); while the latter assumes, as a datum guaranteed by both reason and experience in the scientific method, the *actual inseparability* of the phenomenon and the noumenon, and finds them to be not only compatible, but co-existent and necessary, predicates of the universe *per se*. And the ultimate origin of this fundamental difference lies in the difference between *the subjectivity and the objectivity of relations*, as the only two possible forms of the Theory of Universals, upon which must rest at last the Theory of Knowledge.

According to noumenism, therefore, the noumenon is Intelligible Being, the *mundus intelligibilis;* the phenomenon is Apparent Being, the *mundus sensibilis;* and these two are different yet entirely compatible conceptions of the one universe *per se* which is actually known by science. Phenomenism, being essentially an affirmation of the incompatibility of Real Being and Ideal Appearance, is the victim of the false opposition between the two which the Kantian philosophy derived from mediæval Scholasticism; and philosophy can never become truly modernized until it discards phenomenism altogether, thereby ridding itself of the numberless contradictions latent in this mistaken theory. Restore the true opposition between the phenomenon and the non-phenomenal; restore the Aristotelian principle of the necessary inseparability of the phenomenon and the noumenon; restore the universal Greek principle, unconsciously assumed rather than consciously comprehended and critically justified, of the objectivity of relations; add to these the incontrovertible discoveries achieved by the scientific method in consequence of its adoption of these very principles,—

THE PHILOSOPHY OF SCIENCE. 101

and the whole of modern phenomenism collapses
with its cause, philosophy revives, and man is once
more at home in a universe which he can increas-
ingly know. •

§ 28. For whatever exists is intelligible, because
it is or may be apparent; only Non-Being is unintelligible, because it must forever remain non-apparent.
There are, and can be, no *unintelligible* things-in-
themselves:" so far phenomenism is unquestionably
right. But things-in-themselves are necessarily intelligible: and so far phenomenism is as unquestionably wrong. So understood, the dictum of Hegel
would be true: "Whatever is real is rational."[1]
There exists no "Unknowable," Spencer to the contrary notwithstanding; the only "Unknowable" is
the non-existent. Human intelligence is a light in
the midst of a boundless darkness; its rays shoot
indefinitely far in all directions, and its brightness
grows, fed by a marvellous internal source of illumination whose limits have never yet been ascer-
tained. Whoever presumes to set impassable bounds,
whether deduced from the nature of the darkness
per se or from the nature of the glimmering light
per se, to the area over which it may shine, is guilty
of that worst vice in philosophy — dogmatism, or
the conceit of knowledge without the reality. Increase the light infinitely, and it would expel the

[1] " Was vernünftig ist, das ist wirklich ; und was wirklich ist, das
ist vernünftig." (Hegel, *Werke*, VIII. 17. In his *Werke*, VI. 10,
Hegel himself makes a mistaken reference to this passage, quoted
from himself as " S. XIX." instead of p. 17.)

infinite darkness: the only reason why the infinite
darkness is not absolutely expelled by the light of
human intelligence is that the light is so small.
The existence of the Unknown is. a legitimate inference from the fact of the constant increase of
human knowledge; but to affirm the existence of
that which is *per se* the "Unknowable" is to affirm
and deny knowledge of it in one and the same breath;
and, of all dreary inventions of human pedantry,
Agnosticism is the dreariest, when it elevates this
self-destructive concept of a Known Unknowable into
a mock deity, and founds upon it a mock religion. Is
it not time to lay this "Cock-lane Ghost" of the
Unknowable, and return to the grand seriousness
and simplicity of Greek objectivism?

§ 29. From all this it follows that phenomenism,
on the one hand, is founded upon the *Subjectivity
of Relations* and the *Separability of Noumenon and
Phenomenon;* while noumenism, on the other hand,
is founded on the *Objectivity of Relations* and the
Inseparability of Noumenon and Phenomenon.

§ 30. This last principle is involved in the bare
definitions of the words phenomenon and noumenon,
as respectively "that which is apparent" and "that
which is knowable or known." That which is apparent must be so far known; that which is known
must be so far apparent. Consequently, noumenon
and phenomenon reciprocally contain each other;
they are merely different determinations of *that
which is;* and these determinations are as insepa-

rable as color and form in an object of vision. Whatever appears must exist; the phenomenon without
the noumenon is at once an impossibility and an
absurdity.[1] The case of dreams, hallucinations, in-
sane delusions, and so forth, occasions no difficulty
whatever, for nothing is ever the object of an illusion
which has not, at least in its separate elements, been
noumenally as well as phenomenally experienced.
The dream or delusion, therefore, in no wise differs
from the picture created by the sane waking imagination, except that the dream-synthesis is not, as is the
case with the picture-synthesis, regulated by the intellect. A false appearance is no real appearance ;
by the very terms of the hypothesis, it is false, unreal,
ideal only, — not *Erscheinung*, but *Schein*. What dis-
tinguishes appearance from apparition or delusion,
Erscheinung from *Schein*, is congruity with the en-
tirety of experience ; there is no positive test of
knowledge or criterion of truth save universal human experience, which constitutes the final appeal
of science itself.

But appearance may be either real or ideal. Real
appearance is the appearance of the noumenon-object
in experience ; ideal appearance is the appearance of
the noumenon-subject in consciousness; in either
case, noumenon and phenomenon are inseparable,
and the phenomenon depends upon the noumenon,
since every appearance must be of that which is

[1] " Was erscheinen soll, muss als seiend vorausgesetzt werden."
(Krug, *Lexikon*, I. 835.)

both existent and intelligible. If, as happens in delusions, ideal appearance in consciousness is mistaken for real appearance in experience (and this is the whole fact covered by the expression "false appearance"), what truly appears is the noumenon-subject, disordered in its functions and disguised from itself; the mistake is a mistake of inference as to causation, a wrong interpretation of facts, and, if curable at all, is to be cured by the appeal to universal human experience. Consciousness is always a part of experience, but only a part of it, which phenomenism confounds with the whole.[1] Experience itself, as conceived by noumenism, and as confirmed by science, is the joint product of two equally important factors, noumenon-subject and noumenon-object,— the actual co-existence, union, and inter-penetration of real appearance and ideal appearance, as above defined. Phenomenism misconceives it as ideal appearance alone (*Vorstellung*), and even in this abolishes the noumenon-subject; it thereby irretrievably mangles the fact of experience, *first*, by denying in it the real appearance of the noumenon-object, and, *secondly*, by denying even in the ideal appearance the existence of the noumenon-subject.

[1] "Die aus dem genannten Bedürfnisse hervorgehende Entstehung der Philosophie hat die Erfahrung, das unmittelbare und raisonnirende Bewusstsein, zum Ausgangspunkte." (Hegel. *Werke*, VI. 18. Mr. William Wallace, in his *Logic of Hegel*, p. 15, translates this passage as follows: "The first beginnings of philosophy date from these cravings of thought. It takes its departure from Experience; including under that name our immediate consciousness and the processes of inference from it.")

Science, as actual knowledge of the universe *per se*,
is demonstration of the fact that real appearance of
the noumenon-object and ideal appearance of the
noumenon-subject are actually welded or fused together in experience. Experience is the chemical
union, so to speak, of the noumenon-object and the
noumenon-subject, the former appearing really and
the latter appearing ideally, in a positive third which
is neither one nor the other of the two elements
alone, but a positive coalescence of both essentially
different from either; objective existence and subjective consciousness meet in an actual relation of
action and reaction, or mutual co-activity, which
constitutes the relation of *human knowledge* — the
actual empirical unity of knower and known, subject and object, Thought and Being. Hence all
human knowledge arises in experience; in all experience the activity of Being is the logical *prius*,
and that of Thought the logical *posterius ;* and the
Kantian assumption of "pure *à priori* knowledge"
falls to the ground. In other words, consciousness
itself originates only in experience, and experience
originates in the influence of that which can be
known upon that which can know; but that which
can know must exist before it can be influenced, and
is so far truly *à priori*.

§ 31. Noumenism thus repudiates the fundamental
dualism which compels phenomenism to set noumenon and phenomenon, being and appearance, substance and quality, over against each other as not

only distinguishable in thought (which they are),
but also as separable in fact (which they are not),
and grounds itself on the fundamental monism which
posits the objective identity and merely subjective
difference of the two. Phenomenism grossly carica-
tures noumenism, when it makes the latter conceive
the noumenon as a mysterious and incomprehensible
" substratum " from which phenomenal qualities can
be peeled off one by one, like the coats of an onion ;
and it wins a cheap enough victory over a man-of-
straw antagonist, when it triumphantly inquires what
is left of the onion when the coats are all gone. The
ground of this absurdity lies in phenomenism itself,
not in noumenism ; for it is the former, not the latter,
which assumes the separability of noumenon and phenomenon, — it is the former, not the latter, which
detaches phenomena from the world of Being, trans-
fers them to the world of Thought, and thereby
reduces the noumenon to nonentity as an impossible
" substratum." Noumenism, on the contrary, vetoes
the first step in this royal progress towards nonsense,
and maintains the absolute inseparability of noumenon and phenomenon, — characterizes it as the
quintessence of unreason even to suggest that substantial Being can possibly or imaginably be stripped
of all or any one of its qualities, or that its qualities
can possibly or imaginably be transferred to Thought.
The inherent changeableness of phenomena is a fact
which militates against noumenism no more than
against phenomenism ; for, on either theory, phe-

nomena constantly change. All phenomena, however, must either inhere in noumenal Being as their
ultimate origin and ground, or else must originate
de nihilo and return *in nihilum*.[1] To phenomenism,
therefore, their constant changes are utterly inexplicable, because conceived by it as utterly without
origin — that is, as absolutely ultimate facts; while
to noumenism they are at least partially explicable,
because conceived by it as effects or self-manifesta-
tions of causative or self-manifesting Being, permanent and one. So far as recognition of the Many is
concerned, therefore, phenomenism and noumenism
stand on precisely the same level; but, so far as
recognition of the One is concerned, noumenism
possesses an immeasurable philosophical superiority,
if philosophy is indeed a search for the One in the
Many.

§ 32. Noumenism, then, conceives the universe
as, at the same time, noumenal and phenomenal
both. It revives, though in a far higher form, the
ancient Greek principles of the objectivity of relations and the inseparability of noumenon and phenomenon, and finds the noumenal or intelligible
character of the universe *per se* to consist in its
Immanent Relational Constitution. It beholds in the
modern scientific method the perfection or culmination of actual human experience, the source of all

[1] "——gigni
De nihilo nihilum, in nihilum nil posse reverti."
(Persius, *Sat.* III. 83, 84.)

actual discoveries of truth, and the "promise and potency" of illimitable discovery in the future. It concedes the claim of science to have already discovered an "objective synthesis" of relations in the universe *per se*, existent not merely when they are perceived by man, but just as much in the intervals of his perception ; and it not only repudiates, but reprehends, the essentially strategical policy of phenomenism in misrepresenting and belittling this "objective synthesis" of cosmical relations as a mere "subjective synthesis" of human representations — a policy which proves that phenomenism stands in need of either a little scientific illumination or a little ethical instruction.

§ 33. Further, noumenism argues that, if science has succeeded in discovering objective relations in the universe *per se*, totally independent for their existence on man, his representations, or his consciousness in any sense of the word (and it is an inexcusable belying of science to say of it anything less than that). then there must be in the human mind some adequate and appropriate intellectual faculty, or function, by which they have been discovered. It argues that, since the noumenal universe is actually known by man (the results of science being the self-evident proof of that fact), there must be in man a *Perceptive Understanding* capable of apprehending these indisputably discovered objective relations. It ·is not practicable in this connection to do more than barely touch on this highly important subject, which is developed further in § 50. Enough to say here that what philosophy and psychology have to do, as the most urgently needed service they can render in the present condition of thought, is, not to deny the undeniable as phenomenism does, but patiently to exert their utmost ingenuity to investigate and discover what this now unrecognized mode of knowing is.

' § 34. The main positions of the theory of noumenism may now be presented synoptically in the following summary : —

1. The universe is both a noumenon and a phenomenon, indissolubly one.

2. It is a noumenon because it exists and is intelligible in itself (*per se, an sich*), independent of, yet knowable by, the human mind; and its knowableness or intelligible character consists in its immanent relational constitution.

3. It is a phenomenon because it is apparent and actually known, in part, not in whole; and science is the knowledge of it.

4. Every phenomenon is necessarily a noumenon, and every noumenon is an actual or possible phenomenon. The actual phenomena of the universe constitute the Known ; the universe *per se* is known so far as it is actually related to man's consciousness. The merely possible phenomena of the universe constitute the Unknown; the universe *per se* is unknown so far as it is potentially — that is, not yet actually — related to man's consciousness. But, inasmuch as

no reason is discoverable, either in human consciousness or in the universe *per se*, why the sphere of this
actual relationship may not be indefinitely extended,
and inasmuch as all noumena are necessarily *per se*
intelligible, the Unknowable *per se* is a figment of
imagination which is intrinsically self-contradictory,
and therefore an offence to reason, unless it is conceived as the Non-Existent or the Nonsensical.

5. The human mind includes a perceptive understanding, by which the relational constitution of
the universe *per se* has been already, to some extent,
discovered and formulated in the propositions of
science.   Its function is to apprehend the particular
objective relations immanent in the universe *per se*,
so far as they are presented to human consciousness.
Consequently, the concept of experience must be so
far enlarged as to include, not only the activity of
the senses, but also the activity of the perceptive
understanding (intellection, intellectual perception
or apprehension or intuition).   Science has thus had
a strictly experiential origin, and been built up by
means of that *à posteriori* knowledge of noumena of
which Kant merely assumed, without proving, the
actual impossibility.

This theory of noumenism is nothing but the
logical development of the philosophical presuppositions which were presented at the outset as scientific
realism.   It has been worked out, both in general
scope and special detail, far more than can here be
even hinted.   But enough has been said to show

that modern science contains, lying latent in its hitherto empirical " scientific method," a whole philosophy; and that the stability of all its results, as the " objective synthesis" of a universe which is not the product of man, but the producer of man, must depend in the last analysis upon the soundness of that philosophy. Whatever influence modern science may be to-day exerting on the religious thought of mankind, and whatever influence it may hereafter exert, must proceed, not from the single sciences as such, but solely from the possible philosophies which men may imagine to underlie them as a whole; and the philosophical students of this nineteenth century must be blind indeed, if they fail to see the incalculable importance of developing this necessary scientific philosophy according to true and just principles. The single sciences as such conduct to no universal philosophical conclusion; and for this reason scientific specialists are confident in protestation that " science has nothing to do with religion." But the sciences as a whole, above all the universal scientific method which has produced them, constitute the only foundation on which the philosophy of the future can be reared; and if, as I profoundly believe, human thought is the architect of all things human, then what the philosophy of the future shall prove to be, that also will be its religion.

§ 35. The appended tables, epitomizing the results of the first three chapters of this little book, will conveniently exhibit the relations of the theory of noumenism to the theory of phenomenism and to the history of philosophy in general.

The first table, in particular, will render clearer the general argument of §§ 22–26, and explain the proximate historical origin in Kantism of the phenomenist principle of the separability of noumenon and phenomenon; while the second and third tables will facilitate comprehension of the profound and irreconcilable differences between modern science and (so-called) modern philosophy. If a sharp issue is the necessary condition of every important advance in knowledge, these tables will well repay careful study.

THREE TABLES

ILLUSTRATING

THE ANTITHESIS OF PHENOMENISM AND NOUMENISM.

———•———

I. KANT'S TWO OPPOSITIONS.

1. TRUE OPPOSITION.

PHENOMENON *versus* NON-PHENOMENON.

THE APPARENT *versus* THE NON-APPARENT.

("*Die Erscheinung versus das Nicht-Erscheinende.*")

2. FALSE OPPOSITION.

PHENOMENON *versus* NOUMENON.

IDEAL APPEARANCE *versus* REAL BEING.

("*Die Erscheinung versus das Ding-an-sich.*")

3. HENCE, IN THE KANTIAN SYSTEM,—

NON-PHENOMENON = NOUMENON.

("*Das Nicht-Erscheinende = Das Ding-an-sich.*")

8

II. PHENOMENISM.

MEDIAEVAL NOMINALISM: GERMAN SUBJECTIVISM: MODERN PHILOSOPHICAL IDEALISM.

"Apriorismus."

----

1. GROUND-PRINCIPLE OF THE THEORY OF UNIVERSALS: SUBJECTIVITY OF RELATIONS.

HENCE —

2. GROUND-PRINCIPLE OF THE THEORY OF KNOWLEDGE: SEPARABILITY OF NOUMENON AND PHENOMENON.

Immanent Method = Analysis of Subjective Representation.

RESULTS.

NOUMENON = OBJECTIVELY UNRELATED AND UNINTELLIGIBLE REAL BEING = NON-BEING.
(*"Das Nicht-Erscheinende = Das Ding-an-sich."*)

PHENOMENON = IDEAL APPEARANCE OF SUBJECTIVELY RELATED AND INTELLIGIBLE REPRESENTATION.
(*"Die Erscheinung = Die Vorstellung."*)

III. NOUMENISM.

GREEK OBJECTIVISM: MODERN RELATIONISM:

MODERN SCIENTIFIC REALISM.

"Aposteriorismus."

———

GROUND-PRINCIPLE OF THE THEORY OF UNIVERSALS:

OBJECTIVITY OF RELATIONS.

HENCE —

2. GROUND-PRINCIPLE OF THE THEORY OF KNOWLEDGE:

INSEPARABILITY OF NOUMENON AND PHENOMENON.

———

Scientific Method = Analysis of Objective Experience.

———

RESULTS.

NOUMENON = OBJECTIVELY RELATED AND INTELLIGIBLE REAL BEING = IMMANENT RELATIONAL CONSTITUTION OF THE THING-IN-ITSELF.

PHENOMENON = REAL AND IDEAL APPEARANCE OF OBJECTIVELY AND SUBJECTIVELY RELATED AND INTELLIGIBLE REAL BEING = REAL AND IDEAL APPEARANCE OF THE NOUMENAL THING-IN-ITSELF.

PART II.

THE RELIGION OF SCIENCE.

PART II.

THE RELIGION OF SCIENCE.

CHAPTER IV.

THE PRINCIPLES OF SCIENTIFIC THEISM.

§ 36. WHAT, then, must be the religious outcome of the philosophy logically presupposed by, or latent in, the universal Scientific Method ?

For more than twenty years I have tried to peer into the obscurity of the future and discern the large outlines of this religious philosophy fated to come.[1] I have sought to discover them, not by the comparatively superficial process of forming merely a "widest generalization," which is simply detecting more comprehensive relations in already won *scientific results*, but by going back and down to that underlying *scientific method* which is the creator of all these results, pondering its deeply hidden and fundamental presuppositions, drawing out its subtile implications, and penetrating into the interior recesses of its all-pervading spirit. For the scientific method itself is

[1] See article on "Positivism in Theology," published in *The Christian Examiner*, Boston, March, 1866.

the grandest discovery yet made by man, towering immeasurably above all his other achievements; it is the mother of all achievements, all investigations, all discoveries, — nay, exists immanently in them all as their innermost process and law, and gives them all their meaning; it is man's nearest approach to that secret laboratory of Nature whither her marvellous constructiveness must be tracked back to its birthplace in the eternally creative unity of Being and Thought. The issue of this long meditation has been the "philosophy of science" of which only a few of the most prominent features have been sketched in Part First, yet enough, I trust, to give some conception of the groundwork of that mode of viewing the universe, that *Weltanschauung*, which remains to be unfolded as my anticipation of the "religion of science."

§ 37. Grasp that conception clearly. All Being is essentially intelligible, and either is, or may be, apparent. The Known is actually apparent Being; the Unknown is potentially apparent Being; the unity of the Known and the Unknown is Infinite Being, which comprehends them both. The "Unknowable" is nothing but Non-Being — the Non-Existent and the Nonsensical.[1] The pretended "consciousness of the Unknowable" is nothing but the consciousness of our own finitude, — of our own depressing failure, our weariness, sadness, and pain,

---

[1] "But nonsense never can be understood." (Dryden, *Hind and Panther*, Part I.)

when we strive to comprehend Infinite Being in its totality with our intrinsically finite powers, — of our own bewildered and half-terrified shrinking back into ourselves, when we consciously confront the awful and overwhelming mystery of the Unknown. Sound dies beyond the boundary of our little atmosphere; sight fails beyond the horizon of our little field of vision; thought itself expires in the boundless vacuity of the Unrevealed. But nowhere in Being is there any positive barrier to stop the slow and gradual extension of human Knowledge. Of all forms of dogmatism, the most abhorrent to a sound, sane, and vigorous intellect is the presumptuous audacity which dares to set up flimsy *à priori* "limits of knowledge," or Romulus-walls, to be at once overleaped with a laugh by the Remus of Science, and which, if it only could, would slay him for the deed.

§ 38. However narrow may seem the territory which science has already won from the " void and formless infinite," it is immeasurably vast, compared with the actual or possible acquisition of any individual; and it is real in the highest conception of reality. The ground on which we stand is honest and stable ground, no treacherous quicksand threatening to engulf us if we stir hand or foot, — a tiny floating island in the ocean of the infinite, if you please, but an island every whit as real as the ocean itself. Science maintains that the universe it knows is actual existence, perish who or what may, —

affirms the uttermost reality of its own conquests, — claims to have solved by victorious wit not a few of the Sphinx-riddles propounded to mankind by the *Weltgeist*, — and testifies that it finds the universe intelligible wherever it can bring to bear its unfailing method of research and discovery. It indignantly spurns the sophistry which would explain away its hard-won cosmical truths as the phenomenist's merely subjective "representations" — real while he wakes, potential only while he sleeps. It refuses this proffered kingdom of man's dreams, and vindicates for itself a higher office than merely to introduce into his little phantasmagoric world the coherency, connection, and order which it is laboriously discovering in the universal world of Nature. Nature herself is what science explores and studies, not the mere domain of human "representation;" consciousness is the means it uses, but knowledge of consciousness is not the end it seeks and attains. The phenomenist, who, reversing the precedent of the Hebrew legend, imagines himself to have swallowed the universe, or who escapes the somewhat awkward immodesty of this assumption by sharing the glory of the feat with a host of fellow-phenomenists, is shut down, however reluctantly, to this dilemma: either science is all one huge illusion, or else consciousness is able to apply itself to that which exists beyond its own limits, and to discover in the noumenal world relations which there exist in total independence of that which merely discovers them.

The inconsequent phenomenism which shrinks from this dilemma is entitled to no serious consideration: consequent phenomenism is the pure negation of science.

§ 39. It is no *à priori* assumption, resting on contempt for experience or the rashness of over-confident speculation, to hold that the universe is intelligible through and through, whether within or without the confines of actual human knowledge. On the contrary, this conclusion is a pure induction from experience itself, and the absolutely strongest induction which experience can yield. For every discovery, nay, every perception, ever made by man from the very birth of human intellect, has been a *conversion of the unknown into the known,* — a demonstration, therefore, that the unknown is intrinsically knowable. All knowledge is acquired gradually, or *learned;* and "to learn" is itself to convert the unknown into the known. The totality of human experience itself, therefore, the entire experience of all men in all ages and all climes, is the foundation of this overwhelmingly convincing induction that the unknown is knowable *per se.* What other truth won by man can boast a warrant more absolute? This undeniable *Knowableness of the Unknown,* this experientially proved *Intelligibility of Infinite Being,* is a fact in which there is unspeakable courage and hope for the truth-hungry thinker, who, when the grin-without-a-cat theory assures him that his consciousness can never know anything that depends not for existence on itself, has all human experience on his side, when he replies: "The universe depends not for existence on my consciousness, but my consciousness on it; science has taught me much of it already; and philosophy is an impostor, if she cannot tell me how, and help me to learn more."

Dream as phenomenism may, the fact stands firm, if there is any firmness in modern science and the modern scientific method, that the universe *per se* is independent of man, yet thoroughly knowable by man as far as man has wit to know it. Make his wit infinite, and he would know it all. The universe in itself — all there is of it, and it is the All — is intelligible through and through. There is boundlessly much that man does not yet know, but absolutely nothing that cannot in itself be known. In every phenomenal experience, both he and the universe noumenally appear, — he as the noumenon knowing, and it as the noumenon known; for (as has been shown in Part First) the noumenon necessarily exists in the phenomenon. Phenomenism, therefore, reduces itself to mere gibberish, mere "sound and fury signifying nothing," when it takes the Appearance wholly away from the universe and puts it wholly in consciousness, thereby annihilating the very experience which it assumes to explain. Hence the doctrine of the "Unknowable," which has no foundation whatever except the theory of phenomenism, is the concentrated essence of unreason, if made itself the foundation of a philosophy;

and, if this philosophy founded on nothing is then made the foundation of a religion, it becomes thereby the concentrated essence of superstition — the worship of the Non-Existent and the Nonsensical. The *Knowable Unknown* is one thing; the *Known Unknowable* is a very different thing. The former is the doctrine that what is now unknown may yet become known, and is therefore knowable in itself; it is the strongest possible induction from experience. But the latter is the doctrine that what is unknowable in itself is now known; it is the strongest possible contradiction in terms. In short, the Known Unknowable is an absolute myth, and the Agnosticism founded upon it is a parvenu mythology.

§ 40. Noumenism, therefore, or the philosophy latent in the modern scientific method, establishes the fundamental principle that self-existent Being, whether known or unknown, is absolutely and infinitely knowable, — that the universe *per se* is intelligible through and through, and transparent to finite thought just as far as finite thought can go. This great principle of the *Infinite Intelligibility of the Universe* is the corner-stone of Scientific Theism ; and its warrant is universal human experience, purified, consolidated, and organized in the scientific method.

§ 41. Few scientific specialists, I admit, show any philosophical comprehension of their own method; but this is the fault, not of their method, but of their specialism, and it will cure itself in time. Those scientific men who possess a native largeness of thought too marked to be belittled and defeated by the cramping tendency of exclusively particular investigations will be the first to welcome a philosophy which shall frankly and consistently ground itself on the scientific method, and prove itself to be a truly faithful interpreter of the scientific spirit. To such as these, the doctrine of the infinite intelligibility of the universe in all its boundless extent will be neither more nor less than that of the objectivity and discoverability of all natural truth; and it will be seen to be what it is — the philosophical confirmation and justification of their already practical conviction that all scientific knowledge is genuinely objective experience of a universe not dependent for its existence on the mere continuance of perception. Their not undeserved contempt for "metaphysics" will then be restricted to the baffling and sterile philosophy which identifies all scientific knowledge with mere subjective representation, and thereby extinguishes the possibility of knowing a real external world.[1] Noumenism maintains the *infinite intelligibility*, phenomenism the *infinite unintelligibility*, of the universe *per se*. Between these two

[1] "Die Art der Beweise ist es, welche dem naturwissenschaftlichen Denker jenen instinctiven Widerwillen gegen die Philosophie einflösst, jenen Widerwillen, der sich zu unserer Zeit, wo auf allen Gebieten des Lebens der Realismus über den Idealismus triumphirt, bis zur souverainen Verachtung gesteigert hat." (Von Hartmann, *Philosophie des Unbewussten*, I. 9, ed. 1882.) Von Hartmann himself takes for his motto: "Speculative Resultate nach inductivnaturwissenschaftlicher Methode."

principles there is no logical or rational middle
ground. Which of the two is the more faithful
interpreter of the scientific method and spirit?
Large-minded men of science, especially those of
the rising generation who have escaped the subtile,
contagious, and widespread influence which phenomenism exerts even in scientific circles, will have
no difficulty in answering that question, and detecting the sophistry in the phenomenistic use of the
word "phenomena." For by "phenomena" the
theory of phenomenism means only the *ideal appearance of subjective representation*, while by the
same word science means the *real appearance of
objective being;* and scientific men who once understand the profound difference between these two
things will never concede that the laws of Nature
are valid only in the former sense of this much
abused word. Hence it is hardly presumptuous to
believe that scientific men themselves, whether pre-
pared to go with me further or not, will at least go
with me thus far without the slightest hesitation,
admit that noumenism is the only just and philosophical interpretation of the scientific method, and
concede the truth of the principle that the universe
per se, as discovered by the use of that method, is
infinitely intelligible.

Clearly conceiving the universe as noumenism
conceives it, then, and following as guide the fundamental principle of the infinite intelligibility of
Nature, the unprejudiced and thoughtful mind is

led, I think irresistibly, to momentous conclusions. But, before proceeding to apply this principle, it is necessary to determine precisely what we are to understand by "intelligibility," and also by "intelligence."

§ 42. What, then, is "intelligibility"?

§ 43. Strictly speaking, nothing is intelligible but *relations,* which I have already called the specific and only direct objects of the understanding or intellect (§ 24). Now there is no relation but in and with its terms — no relation but in and with the things of which it is the relation. Things and their relations, though necessarily distinguishable, are absolutely inseparable in Being and in Thought. It was the great defect of the old Scholastic Realism to treat relations as if they were things, and conceive them as separate entities; it is the great merit of the new Scientific Realism to treat things and relations as two totally distinct orders of objective reality, indissolubly united and mutually dependent, yet for all that utterly unlike in themselves.

§ 44. The thing (τόδε τι, *hoc aliquid unum numero, das Ding, das Etwas*) is a unitary system of closely correlated internal forces, and manifests itself by specific qualities, actions, or motions; the qualities, actions, or motions constitute it a phenomenon; the system of relations constitutes it a noumenon, — constitutes, that is, both the real unity of the thing and its intelligible character. This immanent relational constitution of the single thing is, according to the

theory of noumenism, the true "principle of indi-
viduation" (*principium individualitatis -- quodvis in-
dividuum est omnimode determinatum*); perception
never exhausts or discovers all the single relations
or determinations which it includes, although pro-
longed attention always discovers more and more
of them; it is never known wholly, which, however,
is no reason for denying that it is known in part by
science. Scientific discovery has thus far stopped
with the *atom* and the *person,* as the practical limits
of its analysis of the universe into single things
(μόναδες, *Einzelwesen, Einzeldinge*); the universe itself is the All-Thing (*Allding*); between these ex-
tremes is a countless multitude of intermediate
composite things (molecules, masses, compounds,
species, genera, families, societies, states, etc.). The
systems of internal relations in all these various
things vary immensely in complexity and comprehensiveness, — in fact, the complexity and comprehensiveness of the system determines the grade
of the thing in the scale of being; but in every
case the *immanent relational constitution* of the thing
constitutes its real unity, quiddity, noumenal essence, substantial form, formal cause, or *objectively
intelligible character.* Notwithstanding the confus-
ing influence of the theory of phenomenism, a more
or less incomplete perception of this profound truth
asserts itself in philosophers of widely divergent
tendencies; as, for instance, Kant and Fichte, on
the one hand, and George Henry Lewes, on the

9

other hand,[1] who all agree in the acknowledgment
that, so far as it is knowable, the thing essentially
consists in its own system of internal relations.
Consequently, it may be taken as a generally conceded truth that nothing is intelligible except relations. And intelligibility itself as an attribute or
predicate of things, may now be defined, in the language of the schools, as the possession of a determinate essential or substantial form — in the language
of noumenism, as *the possession of an immanent relational constitution*, or system of internal relations.

§ 45. From the fact, however, that nothing is
intelligible except relations, it is not to be inferred,
and does not follow, that all relations are intelligible
in themselves alone. There are relations of disorder, discord, or chaos, no less than of order, harmony, or cosmos. Order is life, disorder is death ;
and disorderly relations constitute the possibility of
death. Taken by themselves alone, disorderly rela-

---

[1] "Dagegen sind die innern Bestimmungen einer *substantia phœ-
nomenon* im Raume nichts, als Verhältnisse, und sie selbst ganz und
gar ein Inbegriff von lauter Relationen." (Kant, *Werke*, III. 228,
ed. Hart.) — "In der Form besteht das Wesen der Sache (*forma dat
esse rei*, hiess er bei den Scholastikern), sofern dieses durch Vernunft erkannt werden soll." (Ibid. VI. 480.) — "Alle diese Verhältnisse mit einander sind das Ding." (Fichte, *Werke*, I. 443.) —
"To know a thing is to know its relations: it *is* its relations."
(Lewes, *Problems of Life and Mind*, 1st Series, p. 59, Amer. ed.) —
"The thing is its relations." (Ibid. p. 89.) All these statements,
of course, must be taken, if fairly interpreted, in a phenomenistic,
not noumenistic sense, — that is, as referring only to the things
of purely phenomenal experience. But noumenism extends them
to things as at once both phenomena and noumena.

tions are absolutely unintelligible; they are relations that do not relate, mere undoing of intelligible relationship, mere dissolution of system, mere nonsense; they are the absolute defeat of intelligence, and its only possible defeat; they could not in any wise exist, if they had to exist by themselves alone, for independent existence is necessarily intelligible. But they do not exist by themselves alone, and herein lies the only possibility of their existing at all; they can only exist in dependence upon, and as parts of, a larger inclusive system which is itself intelligible, and in which they themselves become intelligible by ceasing to be relations of disorder. In other words, disorder, discord, or chaos is not possible as such except relatively to the particular system in which it arises; it is not itself relatively to the larger inclusive system in which this particular system is merely a part; it is an incident of the finite alone, and cannot reach to the infinite. For instance, the decay of an organic cell is disorder and consequent death to the system of that cell, yet order and life to the system of the whole organism, since without the incessant disintegration and excretion of its exhausted cells the whole organism could not live and renew itself; and, again, the decay of the whole organism is disorder and consequent death to the system of that organism, yet order and life to the system of animate Nature, since without the disintegration and excretion of its exhausted organisms the system of animate Nature could not live and renew itself. So the ravings of a maniac are nothing but unintelligible disorder to the unscientific listener, yet intelligible and orderly enough to the sagacious physician, who sees that they are to be rightly related, not to the system of common human experience, but to the vaster system of physiologico-psychological laws, which is the condition of the existence of common human experience, and of which even pathological relations are only normal illustrations. Thus relations of disorder, disease, or death, when viewed from a higher standpoint, become relations of order, health, or life, and therefore intelligible. Chaos *per se* is a stark impossibility; cosmos *per se* is alone possible. For chaos *per se* is an absolute unreality, or pure Non-Being; chaos as a relative reality is simply uncomprehended cosmos, or Being as the Knowable Unknown, and is possible only in relation to the finite intelligence which fails to comprehend it. An actual universe can exist only on condition that it be cosmos, and not chaos; for an actual universe must be self-existent, and self-existent chaos would be nothing but self-existent universal disorder — that is, a *self-existent system of non-system*, which is a flat contradiction in terms.

§ 46. Hence our critical examination of the fact of disorderly relations leads once more to results substantially the same as our former results: (1) that no thing could be intelligible, if it did not exist; (2) that no thing could exist, if it were not intelligible; and (3) that no thing could either exist or be intelligible, if it did not possess an immanent relational constitution. To state these results in more general terms: (1) existence is the condition of intelligibility; (2) intelligibility is the condition of existence; (3) an immanent relational constitution is the condition both of intelligibility and of existence — their aboriginal common ground of possibility, and therefore the absolute ground of the identity of Being and Thought. The immanent relational constitution as such, therefore, is seen to be the common or middle term between Being and Thought, — at once the ground-form of all determinate existence and the grand master-key of all philosophy (§ 84). Finally, to apply these results to the problem in hand: *the infinite intelligibility of the universe, as the infinite, eternal, and self-existent All-Thing, lies in its possession of an infinite and immanent relational constitution.* This is the SYSTEM OF NATURE.

§ 47. The next question to ask is: what is "intelligence"?

§ 48. Phenomenism, the philosophical outcome of the Kantian *Kritikismus*, holds that the nature of intelligence must be determined by the *à priori* analysis of the knowing faculty (*Erkenntnissvermögen*), and that the nature of the object of knowledge — the "what is known" — must be determined by the results of this *à priori* analysis. Tennemann has well pointed out that the essential method of

Scholasticism was to draw all knowledge from the
à *priori* analysis of concepts; and Kantism luminously manifests its own genetic derivation from
Scholasticism by this essential method of drawing
all knowledge from the à *priori* analysis of the
conceiving faculty. Noumenism, on the contrary,
the philosophical outcome of the scientific method,
holds that the nature of intelligence must be determined by the à *posteriori* analysis of the object of
knowledge; that the constitution of "that which
knows" can only be learned from the constitution of
"that which is known;" that actual experience is
the sole revealer of either, and in experience the subject is revealed only so far as it actually experiences.
Hence it argues that the question, "what is known?"
comes first in order, and the question, "what knows
it?" or, "how is it known?" comes afterwards.[1]

§ 49. The fact, therefore, that no objective reality
(apart from Space, Time, and Force, the universal

[1] "An act of knowledge is only possible in relation to an object,
— and it is an act of one kind or another only by special relation
to a particular object. Thus the object at once determines the
existence, and specifies the character of the existence, of the
intellectual energy." (Sir W. Hamilton, *Lect. on Met.*, p. 158.)
"It will not suffice for psychology to throw the *onus probandi, e. g.*
the proof that we have a 'faculty' of Intellectual Intuition, on
supporters of the systems of speculation contemplated. The question is one concerning the *contents* of experience, not concerning its
conditions. It will not do to say, — we have no 'organ' for pro-
curing us such and such experiences; we must first inquire what
experiences we actually have, and then will follow the question,
what 'organs' are those by which they are procured." (Dr. Shadworth H. Hodgson, art. on "Philosophy and Science," in the
London *Mind*, Vol. I. p. 233.)

conditions of all reality) is actually known except things and their relations, and that all that is known of the things themselves is their unitary systems of internal relations, — in other words, that nothing is known of the universe *per se* except its immanent relational constitution, — is proof of the fact that the knowing faculty itself, the understanding or intellect, is nothing but the *Faculty of Relations.* Knowing is by no means the whole of human consciousness; neither is the knowing faculty the whole of the human mind. But our present argument does not require an exhaustive psychological classification of the contents of human consciousness or of the functions of the human mind; it limits itself strictly to that which is germane to the matter in hand, and this demands only a brief account of the knowing faculty as such.

The intellect or understanding, then, is that mode of energizing by which the human mind deals with relations. It deals with them in three distinguishable ways, and may be said, therefore, to discharge three distinct functions: (1) perceptive, intuitive, or analytical; (2) conceptive, reproductive, or synthetical; and (3) creative, constructive, or teleological.

§ 50. (1) The perceptive use of the understanding is essentially *intellection* — that is, intellectual apprehension, intellectual observation, intellectual intuition;[1] and its object is always one or more of

___

[1] This last expression, "intellectual intuition (*die intellectuelle Anschauung*)," is used here in a sense substantially identical with

the particular relations which in their totality compose the immanent relational constitution of the
thing in itself. The thing acts upon the mind; the
mind, as sensibility and understanding, reacts upon
itself as affected by the thing, and subsequently, as
will, upon the thing itself; and the result of this
primary action and reaction is the percept, or perception of the thing. The perceptive understanding is always indissolubly associated with sensuous
intuition in perception; the sensibility apprehends
particular unrelated qualities, the understanding apprehends their particular relations; but the two are
necessarily as inseparable in the act of perception
as the two blades of a pair of scissors in the act of
cutting. Inasmuch, however, as only *related* qualities are intelligible, and as all relations of qualities
in the thing belong to its relational constitution, it is
evident that the thing can be *understood*, not by the
sensibility, nor even by the sensibility and under-

its original meaning in Kant, who denoted by it the *à posteriori* cognition of the noumenon by a perceptive or intuitive understanding.
Kant himself, however, in consequence of his assumption of the
exclusive subjectivity of relations, logically enough denied the
actual existence of such a faculty in man, though he admitted its
purely hypothetical existence in possible higher intelligences.
Fichte used the expression to denote the "pure immediate self-
intuition of the I," Schelling to denote the "non-sensuous intuition
of the Absolute as at once a Real-Ideal," and New England Transcendentalists, as Theodore Parker, to denote the "immediate intuition of God." But these mystical meanings of the expression have
no more to do with the precise, strictly limited meaning assigned
to it in the text, than has the earlier mystical νοεῖν or φρόνησις of
Plotinus, or the *experimentum, intellectualis visio,* or *intuitus gnosticus*
of Scotus Erigena.

standing together, but only by the understanding
alone. For instance, the object of vision is formed
color: color (reflected rays of light) is perceived by
the sensibility; form (which is nothing but a system
of relations of outlines, boundaries, or mere limits
of extension) is perceived by the understanding; the
object of vision is perceived, or "seen," far more by
the mind than by the eye, — in fact, is not "seen"
by the eye at all. Sight is the most intellectual of
the senses; in the other senses, the ratio of perception to mere sensation diminishes, in smell almost
to zero. In other words, the pure sensibility is
not an intellectual function of the mind — no part,
therefore, of the knowing faculty.[1]

Now the actually acquired knowledge of the thing
in itself is never exhaustive; the quantity and quality of it are proportional to the power of observation
and degree of attention bestowed upon the thing;
more always remains to be learned. The study of
the thing is the essential work of the perceptive
understanding, which explores the thing's immanent
relational constitution, and discovers more and more
of it, the longer the exploration continues. This

[1] "Ni notre imagination ni nos sens ne nous sauroient jamais
assurer d'aucune chose, si notre ontendement n'y intervient."
(Descartes, *Œuvres*, I. 164, ed. Cousin.) "We may, therefore,
define Intuition as Mental Vision, or as the Perception of Relations." (Lewes, *Problems of Life and Mind*, 1st Ser., p. 341, Boston,
1874. In a footnote, Lewes quotes Whewell as saying in 1849:
"If we were allowed to restrict the use of this term, we might
conveniently confine it to those cases in which we necessarily apprehend relations of things truly, as soon as we conceive the objects
distinctly.")

perceptive exploration of the immanent relational
constitution of the thing in itself is *Analysis;* and
all perceptive use of the understanding is essentially
analytical. Analysis, therefore, succeeds in individu-
alizing the thing, when it has discovered enough of
its immanent relations to render the system of these
relations intelligible as a whole — in a word, when
it has discovered the real unity or substantial form
of the thing. And Analysis itself may now be
defined as *the experiential discovery, made by the
perceptive understanding, of the immanent relational
constitution, or unitary system, of the thing in itself.*

§ 51. (2) The conceptive use of the understanding
is essentially *reproduction,* or the formation of concepts out of the percepts of individual things. The
conceptive understanding unites perceived relations,
after the pattern of the real systematic unity discovered in the thing by the perceptive understanding, into permanent thought-systems, which persist
in the mind after the disappearance of the percepts ;
and these thought-systems, or concepts, it coins into
words, for use in the intellectual commerce of mankind. Words are mere symbols, but concepts are
not symbols at all ; they are relational systems
identical, as far as they go, with the immanent
relational systems of the things analyzed and discovered by the perceptive understanding. Relations
are the common essence of concepts and things ; as
already pointed out (§ 46), "an immanent relational
constitution is the condition both of intelligibility

and of existence, their aboriginal common ground
of possibility, and therefore the absolute ground of
the identity of Being and Thought." Hence there
is no such thing as "symbolical conceptions;" and
the doctrine of symbolical conceptions, of which so
much is made in the "Synthetic Philosophy," is
itself a "pseud-idea." The only truth in it lies in
the fact that all concepts are only partial repro-
ductions of the relational systems of things, only
silhouette likenesses or outline sketches (so to
speak), since, as has just been explained, the work
of analysis by the perceptive understanding is never
exhaustively completed, and the work of synthesis
by the conceptive understanding is therefore incomplete to precisely the same extent. But that this
work is genuinely successful as far as it goes, and
that the relational constitution of the concept is
identical to this extent with that of the thing in
itself, is proved to the satisfaction of the scientific
mind, whenever it receives the corroboration of fresh
experience in scientific verification.

Now concepts are of two sorts: the concept of
the individual thing, and the concept of the kind or
class. The concept of the individual thing is the
joint work of the sensuous imagination and the conceptive understanding, just as the percept of the
individual thing was the joint work of the sensuous
intuition and the perceptive understanding, or intellectual intuition: the sensuous imagination reproduces the sensuous intuition, and the conceptive

understanding reproduces the intellectual intuition; and, in this reproduction of the percept of the individual thing as a concept, the sensuous imagination and the conceptive understanding are just as indissolubly associated in activity, as were the sensuous intuition and the perceptive understanding in its original production. The concept of the individual thing, therefore, may be called the impure concept, the image-concept, or, shortly, the image. The concept of the kind or class, however, is the sole work of the conceptive understanding, and may be called the pure concept, the concept proper, or the universal notion. For, in the individual thing, the object of the understanding is a relational constitution immanent in an actual unity of sensuously perceptible and imaginatively reproducible qualities which is actually presented to perception; whereas, in the kind or class, the object of the understanding is a relational constitution immanent in an actual unity of many individuals as a group or species, which, however, is never actually presented as such to perception. This relational constitution, therefore, cannot be reproduced in an image at all; it is immanent in the group as a group, in the species as a species, but not in the group or species as a strictly individual thing. While, however, the species as such furnishes no percept to sensuous intuition and no image to the sensuous imagination, and cannot, therefore, be a strictly individual thing, it is, for all that, an individual thing of a higher order,

inasmuch as it possesses a relational constitution
immanent in the totality of its individuals as a self-
related whole, although this totality is never presented
to perception in its real unity. Nay, if the species
as a whole, that is, as an assemblage of all the individuals composing it, were ever presented to perception, then it would yield both a percept and an
image ; but, just as the percept would be a percept
of the assemblage, so the image would be an image
of the assemblage, and not of any "generic individual"—which is a sheer absurdity. Hence the
pure concept, or universal notion, is no image at all,
and the puzzle how to form mental pictures corresponding to "general terms" has caused a deplorable
waste of philosophical ingenuity. No such pictures
are possible, not even with the help of that curious
nonentity, the "generic individual." The universal
notion, or concept proper, is a pure thought-system
of relations, reproducing only the objective system
of relations of resemblance among many individuals, — never the image or mental picture of one
individual ; it reproduces the *relational constitution
immanent in the species as a species*, which includes
none of the relations or qualities peculiar to the
individual as an individual; and it is the synthetical
work of the conceptive understanding. Such, also,
is the concept of the abstract quality, the abstract
action, the abstract motion, and so on: all these
are concepts of relations, dropping out of consideration the things related, and capable of still higher

142   *SCIENTIFIC THEISM.*

abstraction as relations of relations, of innumerable grades of remoteness from individual things. In all such cases, the conceptive understanding simply reproduces systems of relations in thought which the perceptive understanding has previously discovered in being. There is no limit to the generalizations, classifications, abstractions, and so forth, which the conceptive understanding may thus permanently fix or *coin* in concepts, words, and definitions.[1] They are all syntheses of real relations, discovered originally by the analyses of the perceptive understanding, and subject to the necessity of verification in

[1] The analogy between the word and the coin is a very instructive one. Man is an immanent relational constitution, or unitary system, of internal forces; and the expenditure of these forces takes the two phenomenally distinct directions of *labor* and *thought.* From the fact of human society arises the necessity of the exchange of labor-products and thought-products, and hence the necessity of symbolic representatives or readily exchangeable measures of them. As symbols, the coin is the measure of labor, and the word is the measure of thought. Language is the money of intellect: unworded thought is its bullion, but worded thought is its exchangeable currency. Thought itself is the original mental wealth; but, if unworded, it is like gold undug in the mine, which is practically useless even to the possessor until mined and minted. In the intellectual commerce of society, words alone are available property. Consequently, whoever is indifferent to accuracy in the use of words is an unskilled laborer in the intellectual world — a trader ignorant of, or indifferent to, the value of the coins he gives and takes; and commercial failure would be the instant fate of him who in business should confound the eagle with the dollar and the dollar with the cent. Hence the vulgar reproach against philosophy that it is mere hair-splitting in words, or profitless trickery in verbal subtilties, is simply a proof of vulgar ignorance. The student of science or philosophy who should despise or neglect the subtile but real distinctions of technical terms would speedily become a scientific or philosophical bankrupt.

fresh experience; they not only perpetuate these discoveries, but also furnish indispensable instrumentalities for the further scientific exploration of objective reality. Hence intellectual reproductive synthesis, or the combination and conversion of evanescent objective percepts into permanent subjective concepts, is the special function of the conceptive understanding.

§ 52. (3) The creative use of the understanding is essentially *teleological*: that is, it is the free construction of *ends and means*. The end is a purely ideal system of relations in the present which is to be realized in the future; the means is a purely ideal system of relations in the present by which the future end is to be realized; and the objective realization of these purely subjective systems of relations is effected by the will, which is the blindly executive faculty or function of the mind and implicitly obeys the directive mandate of the understanding. When the understanding takes the suggestion of its end from feeling, the general end it creates is the attainment of (egoistic or altruistic) happiness, and its principle of action is utility or expediency — that is, fidelity to the immanent relational constitution of the mind itself; when it takes the suggestion of its end from the higher reason (which is the supreme *Faculty of the Ideal*), the general end it creates is the attainment of truth, beauty, and goodness, and its principle of action is justice — that is, fidelity to the immanent relational constitution of

the universe *per se.* There is no necessary antagonism between these two ultimate principles of action;
antagonism arises only when the lower or partial
principle usurps the authority of the higher or universal principle — when feeling asserts an unnatural
and illegitimate supremacy over the higher reason.

The understanding, therefore, and not the will, is
the true *Faculty of Freedom* — intellectual freedom
when the immediate end is the knowledge or application of truth (science, philosophy, and the mechanical arts), æsthetical freedom when the immediate
end is the possession of beauty (literature and the
fine arts), practical or moral freedom when the immediate end is the conduct of life or the achievement
of virtue (praxis, morality, religion). For example,
it creates innumerable objective relational systems
in tools, machines, and other inventions of the industrial arts; pictures, sculptures, musical instruments, buildings, books, all works of the higher
imagination, in the fine arts ; institutions of all sorts,
civil, political, military, philanthropic, ecclesiastical,
in human society; plans of conduct, schemes of
social reform, religious organizations, and so forth,
in the sphere of moral and religious activity, — in
short, all the instrumentalities and enginery of
human civilization.

§ 53. Now, in all this multiform self-activity, the
creative understanding appears as the *absolute originator of systems of relations.* It pervades all other
uses of the understanding, whose functions are dis-

tinguishable only, not separable. The percept, it is true, is a system of relations created by the joint activity of the thing and the understanding, object and subject, noumenon known and noumenon knowing, co-existing and interacting in the phenomenon, real appearance, or actual experience. But the concept is the perpetuation — that is, the *ideal recreation* — of the percept; and the end and the means are combinations of percepts and concepts in *free, absolutely new, and purely ideal creations,* subsequently realizable by the will. Thus the perceptive understanding *discovers* objective systems of relations; the conceptive understanding recreates or *reproduces* them; the creative understanding, in its pure activity, recombines them, and thereby *freely creates* new subjective systems of relations. The supreme construction of the creative understanding is *Method,* which is also the highest perfection of teleology; for it is the adaptation of means to ends, not for a single act or judgment, but for the universal series of acts or judgments. Hence, method being the highest potency of intellect *in actu,* the essentially teleological nature of all intellect is plainly apparent.

In all its functions, the essential act of the understanding is judgment; and judgment is always the affirmation (including, of course, negation) of the objective existence, or fitness to exist objectively, of systems of relations. Reasoning, or the continuous activity of the understanding, is the strictly teleo-

10

logical combination of judgments of objective existence to produce a final judgment of precisely the
same character. The percept is, originally, the judgment or affirmation of the objective existence of particular relations in the thing, and, finally, of their
real unity in its relational system as a whole; the
concept, or the reproduction of the final percept, is
the judgment or re-affirmation of the objective existence of the relational constitution of the thing; the
end is the judgment or affirmation of the fitness of
a purely ideal relational system to exist objectively,
and the means is the judgment or affirmation of the
fitness of another purely ideal relational system to
exist objectively, in order to produce the objective
existence of the end. But this is not all. The
understanding is the absolute originator of systems
of relations, not only in thought, but also in being.
Acting in conjunction with the will as its executive
subordinate, it masters forces which exist in the
outward world, and constrains them to reproduce
relational systems which have absolutely no origin
but the understanding itself. This is the manner in
which the mind, originally acted upon by the thing
in perceptive experience, reacts ultimately upon the
thing itself in teleological construction. For instance,
the ship, as a ship, is the teleological creature of the
understanding alone; its materials and forces are
derived solely from external nature, but its idea, its
real unity as a ship, its immanent relational constitution, without which it would be a mere mass of

timbers and cordage or material still less formed, is derived solely from the understanding, which has absolutely created it as a means to its own ends. The ship, as such, is nothing but this immanent relational constitution, this real unity of relational system, this substantial form ; and, in virtue of this internal system alone, the ship is truly a thing in itself. Hence the ship, as a thing in itself, is a system of objectively real relations created absolutely in the world of actual existence by the understanding and the will. And the general result of our in vestigation of the nature of intelligence may now be condensed into this brief definition, to be interpreted in the light of what precedes : *Intelligence is that which either discovers or creates relational systems or constitutions.*

§ 54. It only remains to show, under this head, that the nature of intelligence, as such, is identical in all possible forms and degrees.

Any organism, however low in the scale of being, which has sufficient intelligence to select its food, choosing the nutritious and rejecting the innutritious, — or to fly from its enemies, or to seek shelter from the weather, or to seek its mate, — proves thereby its possession of a perceptive understanding, or the capability of discovering systems of relations objectively real to itself. Any organism which manifests the ability to act for a purpose or end, however simple, proves thereby its possession, not only of a perceptive, but also of a conceptive and creative, understanding. Many of the lower animals manifest rudiments even of the higher reason, the faculty of the ideal, so far as they show themselves capable of self-sacrifice for the sake of others. So far as modern investigations go, they tend to prove that mind is everywhere mind, essentially identical in kind, however various in degree. If man's understanding were infinite, it could not cease to be what it is, no matter what new faculties might be added to it; it would still be essentially that which cognizes and deals with relations, or it would cease to be mind altogether. The network of relations, the inosculating and interpenetrating systems of relations which in their totality compose the immanent relational constitution of the universe, would still remain to be known; and the infinite mind which should not know them would be simply infinite stupidity and ignorance. Knowledge itself can be nothing but knowledge of these relations; finite knowledge is knowledge of a part of them, infinite knowledge could only be knowledge of them all. There is thus no essential difference in knowledge itself, or in the knowing faculty, whether it be finite or infinite; the difference is in the nature of the object of knowledge, as finite or infinite. Hence man's present intelligence, if only infinitely expanded without the slightest change in its essential nature, would be thereby rendered adequate to the absolute comprehension of the absolute All; and, if there exists anywhere or anyhow an absolute and infinite

mind, its essential nature must be identical with
that of the human mind, differing in degree alone
and not in kind, if, as has just been shown, mind or
intelligence is that which either discovers or creates
relational systems. In only two unessential respects
could an infinite intelligence differ from a finite
intelligence: an infinite understanding would be
perceptive and *creative*, but not *conceptive*; and while,
to the finite intelligence, the material of its perception and creation must be given from without, this
material, to the infinite intelligence, must be given
from within. For, on the one hand, it is only the
non-continuance or evanescence of the percept which
renders the concept a necessity of finite intelligence;
the conceptive understanding is merely a remedy for
the defect or finitude of the perceptive understanding; and an infinitely perceptive understanding
would itself discharge the essential function of the
conceptive understanding, since a permanent percept
would be indistinguishable from the concept and
render the latter superfluous. On the other hand,
the fact that the finite intelligence originally deals
with relational systems only in that which is given
to it from without results likewise from its finitude
as such; an infinite mind would necessarily originate from within both matter and form of the relational systems which, as an infinite perceptive
understanding, it would intuitively comprehend.
Hence there must be unessential differences between
the finite and the infinite intelligence; but the

definition of intelligence itself as " that which either
discovers or creates relational systems " remains
equally applicable to both, and hence the principle
stands unshaken that the nature of intelligence, as
such, is identical in all possible forms and degrees.
If this is " anthropomorphism," it matters not ; hard
names never yet changed a false principle into a
true, or a true principle into a false ; and the question remains as before — is the principle itself false
or true ?   And this principle of the essential identity
of the nature of intelligence in all possible forms
and degrees, following so clearly and necessarily
from the scientific or noumenal conception of the
universe, seems to be undeniably true.

§ 55. Having at last arrived at answers to the
vitally important subsidiary questions, " What is intelligibility ? " and " What is intelligence ? " it is time
to resume the thread of our main argument.   It has
been shown (§ 46) that intelligibility, as an attribute
of the thing, consists in the possession of an immanent relational constitution, and that the infinite
intelligibility of the universe, as the infinite, eternal,
and self-existent All-Thing, consists in its possession
of an immanent and infinite relational constitution.
It has likewise been shown (§ 53) that intelligence
itself is that which either discovers or creates relational systems or constitutions, and that the nature of
intelligence, as such, is identical in all possible forms
and degrees.   What is the unavoidable inference or
conclusion from these principles, as premises ?

This — that *the infinitely intelligible universe must be likewise infinitely intelligent.*

The infinitely intelligible universe is the self-existent totality of all Being, since there is no " other " to which it could possibly owe its existence. But that which is self-existent must be self-determined in all its attributes; and it could not possibly determine itself to be intelligible, unless it were likewise intelligent. Self-existent intelligibility is self-intelligibility, and self-intelligibility is self-intelligence: or, that which intelligibly exists *through* itself must be intelligible *to* itself, and therefore intelligent *in* itself.

To express this thought in less abstract terms: the universe, being the sole cause of its own existence, must be likewise the sole cause of all the determinations of that existence, and therefore of its own intelligibility; that is, it must be the absolute author or eternal originator of its own immanent relational constitution. Intelligence, as the creative understanding, has just been shown to be the " absolute originator of systems of relations;" and no other origin of relational systems is either known in experience or conceivable in hypothesis. So far as experience and reason can go, therefore, the intelligibility or relational system of the universe, considered as an effect, must originate in the intelligence or creative understanding of the universe, considered as a cause. This is substantially the meaning of Spinoza's famous distinction of *natura naturans* and *natura naturata*, and of Prof. Caporali's distinction, identical with Spinoza's, of *natura fatta* and *natura che si fa*. It does not mean causation or creation at any particular time, but the eternal self-causation or self-creation which is involved in the reality of Infinite Being as Eternal Self-activity, *actus purus*, or *causa sui;* and this is a conception which, as we saw in Part First, phenomenism itself is powerless to escape. And it is no less the conception towards which, as pure cosmical dynamism, modern science is steadily tending more and more.

Hence the existence of an intelligible, infinite, and immanent relational constitution or system in Nature is the highest possible or conceivable proof that Nature is intelligent; and the stronger the proof of the system, just so much stronger is the proof of the intelligence. The absolute invariability of natural law, which is the logical corollary of natural system, is thus essential to the conception of an immanent relational constitution as the real unity of the universe; the possibility of miracle, as a suspension of natural law, would be the disproof of an infinite intelligence. Now, as was shown at the outset (§ 4), the scientific conception underlying, or lying latent in, all empirical use of the scientific method is that the universe, as a whole, *has* an immanent relational constitution, and that all the countless particular relations of which it is the real unity are actually or potentially discoverable by observation, experiment, hypothesis, and verification. No scientific investigation could possibly ever have been instituted, except on the conscious or unconscious assumption of the scientific discoverability of these relations *per se:* that is, of the scientific knowableness of the unknown, — that is, of the existence in Nature of a relational system which includes, not only the known, but no less the unknown in all its boundlessness. The whole progress of science, discovering more and more of that intimate relational system which finds place for every new fact as soon as it is brought to light, is a cumulative proof, mounting almost to mathematical demonstration, that the "objective synthesis" or system of Nature is the most real of all realities. Yet this system, as has just been shown, is the strongest possible proof of infinite intelligence in the universe. System — all-inclusive self-relatedness in whole and in part — has, even conjecturally, no possible origin except intelligence ; and an infinite system, inclusive alike of the known and the unknown, can have no origin but in infinite intelligence. Chance, or fate, is no hypothesis at all ;[1] it is the mere absence, the mere negation, of all hypothesis ; intelligence is the only hypothesis in the field, for intelligence, as the creative understanding, is the only experientially known or hypotheti-

[1] "It would be with it as with that man of whom Gassendi speaks, who, half asleep, and hearing four o'clock struck, said : ' This clock is mad ; lo, four times it has struck one o'clock !' The man had not force of mind enough to reflect that four times one o'clock makes four o'clock. Those who explain the world by a fortuitous concourse of atoms give evidence of a power of synthesis about equal to this." (Janet, *Final Causes*, 1883, p. 28.)

cally conceivable origin of system. Yet the system
of Nature is the surest fact of science; it has, and
must have, an adequate cause; and the only cause
known to man which is ever the originator of system is the creative understanding. All experienced
systems which cannot be referred to the creative
understanding in man and the lower animals belong
together, as mere parts of the one total system
known to exist as the immanent relational constitution of the universe.

It will not do to say that man discovers a multiplicity of relational systems in Nature which have
evidently no origin in an originating understanding,
and that system as such, therefore, is no proof of
originating understanding; for not one of these systems is independent — they are all *dependent parts*,
not *independent wholes*, and only constitute elements
in the one vast system of Nature itself, which in its
absolute unity alone explains them or renders them
intelligible. There is one, and only one, System of
Nature; there are many system-products of man,
because man is *many*, but only one system-product
of Nature, because Nature is *one*.

The simple question is — shall this one system, as
a known fact, be referred to anything but intelligence, the known cause of innumerable other known
systems? To this question but one reasonable answer can be given, if experience is the true guide
of reason. Consequently, the immanent and infinite
relational constitution of the universe *per se*, verified

by experience as far as experience has gone, and
confirmed by reason as far as reason can go, is the
one grand and decisive proof that *the infinite intelligibility of the universe can have no possible origin but
the infinite intelligence of the universe itself.*

§ 56. Now let us put two and two together, and
see if they make four. Our results thus far are (1)
that the universe *per se* is infinitely intelligible, and
(2) that the universe *per se* is infinitely intelligent.
Unite these two truths, and the third truth follows
with irresistible certainty that *the universe per se is
an infinite self-consciousness.* For that which is intelligible is an actual or possible object of knowledge; that which is intelligent is an actual subject
of knowledge; and that which in itself is at once
intelligible and intelligent is an actual subject-object
— a living self-consciousness. This actual identity
of subject and object, or "transcendental [experi-
ential] synthesis of Being and Knowing in the I,"
is precisely what constitutes the mystery, and yet
the undeniable fact, of all consciousness (*Urthatsache
des Bewusstseins*). The universe, then, is infinitely
intelligible and infinitely intelligent at the same
time; since it includes all that exists, and therefore excludes the possibility of any other object of
knowledge than itself, it must be its own object; consequently, it must be an actual and infinite subject-
object, that is, an infinite self-consciousness.

Thus far, then, we seem to have been led by a
very straight path, assuming only the validity of the

scientific method and of the philosophical presuppositions logically involved in it, to the momentous result that the universe *per se* is an *Infinite Self-conscious Intellect,* which, though infinitely removed in degree, is yet essentially identical in kind with the human intellect. This result, then, is the constitutive principle of Scientific Theism; and I see no way to escape it, except by repudiating the scientific method itself. But this result is by no means an ultimate one. Let us, then, conclude this long chapter, and in the next go on and see whither the road we are travelling will conduct us.

CHAPTER V.

THE UNIVERSE: MACHINE OR ORGANISM?

§ 57. THE immanent relational constitution of the universe *per se*, then, is the mode in which the universe-subject, or Infinite Self-conscious Intellect, thinks and creates and reveals itself as the universe-object, or infinitely intelligible System of Nature; and, so far as it is yet known, modern physical and psychical science is the knowledge of it. From the side of the finite, science is human discovery; from the side of the infinite, science is divine revelation; there could neither be discovery without revelation nor revelation without discovery; and science thus appears as the intellectual mediator between the finite and the infinite. The philosophy of science, therefore, when at last developed and matured by the universal reason of the race, will be the supreme wisdom of Man and the self-evident word of God. All this seems discouragingly abstract and lifeless; but life and light appear as we go on, following the course of this objectified divine self-thinking in the System of Nature, with science still as our guide.

§ 58. The System of Nature, as the real unity of all existent things in the All-Thing, must be, not

only infinitely intelligible, but also absolutely perfect, in every sense of the word. It must be perfectly adapted to the conditions and laws of Being,
else it could not persevere to be. It must be perfectly
adapted to the conditions and laws of Thought, else
it could not be intrinsically understandable and pro-
gressively understood. It must be perfectly adapted
to itself, perfectly self-related in whole and in part,
perfectly self-constituted as an infinite relational system, else it could neither be nor be understood (§ 46).
Finally, it cannot be imperfect in comparison with
any other and superior system, whether within or
without itself ; for outside of itself there is nothing
with which to compare it, and inside of itself there is
no partial or finite system which does not absolutely
derive from the universal or infinite system whatever
little perfection it may possess. In short, whatever
is imperfect carries in its own imperfection the seed
of death — must at last decay and altogether cease
to be ; but whatever exists eternally proves its own
absolute perfection by the bare fact of its eternity.
Hence the System of Nature must be absolutely
perfect in every conceivable sense of the word.

§ 59. The conception of the universe, therefore, as
nothing but "an infinite multitude of sentient beings" (monads, monadology), or as nothing but "an
infinite multitude of non-sentient beings" (atoms,
materialism), is a distinctly inferior and imperfect conception, and, consequently, cannot correspond with the
fact. It is not one system at all, but an unintelligible

aggregate of systems. It is a conception intermediate between the conceptions of cosmos and chaos, infinite order and infinite disorder, Being and Non-Being; it posits the objective existence of particular relations, but abrogates that of the total relational system through which alone they could objectively exist; it establishes the Many, but abolishes the One; it lacks all principle of real unity, and therefore all principle of self-existence; it lacks all principle of ideal unity, and therefore all principle of intelligibility. By thus destroying all real and ideal unity of the universe, it represents the universe as so imperfectly self-related that, as a system, it would be immeasurably inferior to each and every one of the monadic or atomic systems which are contained within it, and which, notwithstanding, must derive their higher perfection from this universal system less perfect than themselves; for, although the universe, as a whole, is not conceived as self-existent or intelligible, each monad or atom is necessarily so conceived, and thus the part is conceived as superior to the whole. The conception itself, therefore, is essentially a hybrid conception, a cross between cosmos and chaos, a philosophical chimera, a monstrosity, and dissolves the complex unity of the universe into a mob of disorderly elements. As a perfect system, Nature must be, not an infinite multitude of self-existent units, forever clashing and colliding in a turmoil at once hopeless and eternal, but an infinite relational constitution, in which not only the infinite

multitude of the units, but also the infinite unity of
the multitude, are, logically and ontologically alike,
reconciled and conserved.

§ 60. The universe, then, is a self-existent, infinitely intelligible, and absolutely perfect system.
But what is a perfect system ? A system, in general, is that in which many parts are correlated
closely enough to constitute a rational whole; and,
purely as system, it is more or less perfect in proportion to the closeness, complexity, and comprehensiveness of this internal correlation. There are
countless grades of perfection in the finite systems
known in human experience ; each may be perfect in
itself, inasmuch as it may be perfectly adapted to its
own immanent end and its own place in the general
whole, yet at the same time relatively imperfect,
inasmuch as the degree of closeness, complexity, and
comprehensiveness in its internal correlation may be
greatly inferior to that of other finite systems. Moreover, it is incredible, in the light of human experience
itself, that the vast and limitless Unknown should
not conceal from man's perception countless grades
of perfection in systems as yet unrevealed to his
prying eye and mind. But, so far as his knowledge
goes, the supreme perfection of system is realized in
that system of systems — the *Organism*. All other
known systems are immeasurably less perfect than
this, because the organism *lives and grows*. Nothing
but the organism either lives or grows; the knowledge of life and growth is derived from it alone; life

and growth are its essential marks, and constitute it, within the sphere of human knowledge, the one perfect system.

§ 61. Now, in our analysis of the knowing faculty and of the nature of intelligence, we found that the supreme function of the understanding was the teleological creation of system. The creative understanding of man, however, is powerless to create the organism, or one perfect system; it cannot project itself into the world of external existence in any higher form than that of the *Machine,* or relatively imperfect system, because it deals only with given material over which it can exercise only a limited control. Even its highest ideal creations are never emancipated from dependence on the merely given; all human knowledge is drawn from experience of the given Outward, and all human construction is mechanical recombination of the material it yields. The fine arts themselves are only members of the great sisterhood of the mechanical arts : the statue, the painting, the orchestra, the cathedral, nay, the book, are only machines for producing certain effects in the human mind. The industrial àrts minister to the wants of the body ; the fine arts minister to the wants of the mind ; but both are simply departments of the mechanical arts, and equally ultimate in the production of machines.

Hence the finite understanding can create innumerable mechanical or artificial systems as means for the enlargement of its own life, but never organic

systems as means for the creation of life in itself.
If it ever should, it would only prove itself more
divine than it seems. But the infinite understanding, which creates both the form and the matter of
its own constructions, creates organisms, and, rightly
interpreted (as will appear below), organisms alone.
It was a profound saying of Strauss, essentially identical with Aristotle's doctrine of the ἐντελέχεια, that
"life is an end that creates its own means from within and realizes itself."[1] The Infinite Self-conscious
Intellect eternally creates the Infinite Organism of
Nature, — that is, the universe as subject (*natura
naturans*) eternally creates the universe as object
(*natura naturata*), — because self-existence or self-
life is eternally a self-sufficient end that realizes
itself, an end in itself that is not a means to any
further end ; and it creates finite organisms because
even dependent life is likewise, at least in part,
a self-sufficient end that realizes itself. In other
words, life, whether infinite or finite, is its own
justification : you fulfil your " being's end and aim "
by living your own life in all genuineness and ideal
fulness — by truly fulfilling, that is, " full-filling " it ;
and you are wise indeed, if you know the boundless depth of meaning and the vastness of universal

---

[1] "Das Denken kann in diesen Forschungen nicht eher zur
Befriedigung gelangen, als bis es, den ganzen Standpunkt dieser
ausserhalb der Natur entworfenen und ihr eingepflanzten Zweck-
beziehungen verlassend, die Idee des Lebens als den sich von
innen heraus seine Mittel schaffenden, sich selbst verwirklichenden
Zweck begreift." (*Die Christliche Glaubenslehre*, L 388, ed. 1840.)

obligation which the word "full-filling" implies.
The creative understanding, therefore, which is the
absolute originator of all relational systems, creates
them because that is its essential function — because
it is its very nature to create; all its creations are
essentially teleological, — as finite understanding,
machines, and as infinite understanding, organisms;
and all its creations are essentially means for the
"full-filling" of its own life — the absolute and universal end of all Being.

§ 62. Now modern science is rapidly reaching,
nay, has almost reached, this sublime conception
of the universe as a living and growing organism.
Organisms themselves are of countless grades of
perfection. In one sense, every organism is perfect
which is perfectly adapted to itself and its environment; yet organisms so adapted, if considered relatively to each other, are more or less perfect as they
embrace more or less of the environment in those
external relations of their own life which constitute,
as it were, the actual extension of this life. Hence
an organism is higher or more perfect, the more it
projects itself into the outer world, and learns to
subordinate outer forces to its own uses; or, in other
words, in proportion to the strength of its creative
understanding and the consequent effectiveness of
its machines, — that is, its relational systems of all
kinds, created as means for the enlargement and
enrichment, the "full-filling," of its own existence.
This is "judging the tree by its fruit," it is true, but

the test is sound. Man has no better title to his primacy among animals than the potency and vastness of the combinations (relational systems) by which he has mastered natural forces, and practically annexed to his own being so immense a part of the planet he inhabits.

' Now the universe has no environment to master or annex. If, then, it is to be conceived as an organism, it must be conceived as an organism all of whose life and growth are strictly immanent, and different in important respects from the finite and merely individual organism to which the name is usually confined. The finite organism not only lives, but also dies; it lives by drawing into itself, and subordinating to its own uses, that which is not itself, and it dies at last by its inability to convert, absolutely and permanently, this not-itself into itself. But the infinite organism lives, and dies not; it lives by eternally converting itself as *force* into itself as *form*, and it dies not, because it has no need to convert the not-itself into itself — because its eternal. self-conservation is its eternal self-creation.[1]   Again,

---

[1] " En effet, c'est une chose bien claire et bien évidente à tous ceux qui considéreront avec attention la nature de temps, qu'une substance, pour être conservée dans tous les moments qu'elle dure, a besoin du même pouvoir et de la même action qui seroit nécessaire pour la produire et la créer tout de nouveau, si elle n'étoit point encore ; en sorte que c'est une chose que la lumière naturelle nous fait voir clairement, que la conservation et la création ne diffèrent qu'au regard de notre façon de penser, et non point en effet." (Descartes, *Œuvres*, I. 286.)   " Mais il est certain, et c'est une opinion communément reçue entre les théologiens, que l'action par laquelle maintenant il le conserve, est toute la même que celle par laquelle il l'a créé." (Ibid., I. 172, 173.)

the finite organism reproduces itself only by producing another which is not itself, yet like itself;
the form or relational system abides, but is subject
to modification, because the matter changes under
the influence of kindred matter in the environment.
But the infinite organism reproduces itself at every
instant, and does not produce another; its form and
its matter are alike eternal. Again, the finite organism is evolved out of the environment and dissolved
back into the environment. But the eternal evolution and dissolution which constitute the life of the
infinite organism are absolutely immanent within
itself, and do not affect its eternal self-identity.
These differences are important, and should not be
overlooked; they do not, however, touch the essential concept of the organism as that which lives and
grows, and leave it compatible with finitude and
infinitude alike.

§ 63. This conception, then, of the System of
Nature as an Infinite Organism is the highest conception which man has yet formed of the immanent
relational constitution of the universe *per se* — his
nearest actual reproduction in thought of the infinitely intelligible and absolutely perfect system of
universal Being; and it is precisely the conception
which modern science is to-day working out in that
marvellous discovery of the nineteenth century, the
Fact of Evolution. It is true that the law of evolution is not yet successfully formulated, and that the
conception of it has been thus far only imperfectly

developed; neither the formula nor the conception has thus far been philosophically matured in the systems of those who have attempted to philosophize it. In reality, the greatest battle of modern thought turns on the further and profounder determination of the concept of Evolution, and this turns on the determination of the concept of the immanent relational constitution of the universe *per se.* On this great question, phenomenism has absolutely nothing to say; the answer to it lies only in the scientific method, its logical presuppositions, and the theory of noumenism which is the logical development of these presuppositions into a determinate philosophy of science. It is my deep conviction that the final issue of the battle will be the permanent and universally recognized establishment of the conception of the *System of Nature* as an *Infinite Organism.* Science has not yet reached the fulness of this conception, but it lies implicit in the scientific method as the flower lies implicit in the bud, and, whenever it shall have become explicit, science will have become philosophy itself.

§ 64. Now this organic conception of Nature clearly reveals the crudity and falsity of the idea, often broached, that "God comes to consciousness in Man."[1] It is perfectly true that the system of

---

[1] Cf. Hegel, *Werke*, XIII. 48: "Es sind viele Wendungen nöthig, ehe der Geist zum Bewusstseyn seiner kommend sich befreit. Nach dieser allein würdigen Ansicht von der Geschichte der Philosophie ist der Tempel der selbstbewussten Vernunft zu betrachten." On this and similar passages Von Hartmann well

Nature, self-evolved as the objectified divine thought, has risen with incalculable slowness from the unconscious to the conscious; but the whole process remains utterly unintelligible, nay, an absurdity or self-contradiction, unless the evolution of the universe as *Divine Object* is viewed as the work of the universe itself as *Divine Subject,*—that is, as the *Infinite Life of God in Time and Space.* No more can be evolved than is already involved: the conscious could not possibly originate in the unconscious. The notion of "God's coming to consciousness in Man," if it means that no Infinite Self-conscious Intellect existed before man appeared, arises from non-perception of the great principles already explained: namely, that an infinitely intelligible system, as a strictly intellectual effect, can have no origin but an infinite creative understanding, as its strictly intellectual cause,—and that, if infinite intelligibility and infinite intelligence co-exist as eternal attributes in one sole and self-caused existence, as they must in the universe of Being, then that universe must be an infinite subject-object, or Infinite Self-consciousness. Intellect itself is the only known, knowable, or imaginable cause of intelligible system; and Nature, the universal system of

says, *Philosophie des Unbewussten*, I. 23, ed. 1882: "Der Hegel'sche Gott als Ausgangspunct ist erst 'an sich' und unbewusst, nur Gott als Resultat ist 'für sich' und bewusst, ist Geist. . . . Die Theorie des Unbewussten ist die nothwendige, wenn auch bisher meist nur stillschweigende *Voraussetzung* jedes objectiven oder absoluten *Idealismus*, der nicht unzweideutiger Theismus ist."

objective relations, is just as necessarily the product of infinite mind, as philosophy, the universal system of subjective relations, is necessarily the product of finite mind. Hence it is shallow and poverty-struck thinking which conceives that God is originally not infinite self-consciousness, but merely comes to a finite consciousness in man; and which thus fails to see that the evolution of the universe-object, as intelligible system, is explicable only by the universe-subject, as intelligent origin of that system or infinite creative understanding.

§ 65. The organic conception of Nature reveals with equal clearness the crudity and falsity of the idea, also often broached, that "God exists outside of Space and Time." Space and Time are not known at all except as the universal conditions of all existence, — as absolute forms of all Thought because, and only because, they are absolute ground-forms of all Being. Kant's theory of the exclusive subjectivity of Space and Time, as pure *à priori* forms of sensuous intuition, is utterly untenable and self-destructive.[1] The noumenism of the scientific method establishes their necessary objectivity, as *condiciones sine quibus non* of noumena themselves. The universe as divine object, and therefore the universe as divine subject, are thus absolutely conditioned on Space and Time, which, far from being positive determinations or limitations of Being, are only the

[1] See my article on "The Philosophy of Space and Time," in the *North American Review* for July, 1864.

blank forms of its possibility. The attempt, therefore, to deduce Space and Time from God is the
destruction of all intelligibility in the philosophy
which attempts it. In fact, the statement that
"God exists outside of Space and Time" is a double-
barrelled contradiction in terms; for the "exists,"
a verb of present tense, presupposes the very Time
which the "outside of Time" denies, while the
"exists outside" presupposes the very Space which
the "outside of Space" denies. All existence as
necessarily presupposes Time as all matter necessarily presupposes Space; and the statement (if it
had any conceivable meaning) would affirm at the
same time absolute atheism and absolute acosmism,
for, since God and the universe are one, it would
deny all real existence to both in denying it of either.
Not even the phenomenist, however, pretends that
the universe as such "exists outside of Space and
Time;" if he subjectifies Space and Time, he no
less subjectifies the universe, and himself conceives
the former as conditioning the latter in representation or thought. To claim, then, that "God exists
outside of Space and Time" is, on any hypothesis,
at least to banish God from the universe altogether,
and condemn man to be, in the most literal sense,
"without God in the world." But it is a waste of
criticism to expend it on a conception so dismally
chaotic.

§ 66. The fact of evolution, independent of all
theory about it, is to-day established beyond reason-

able doubt as a permanent result of modern science.
The conception of evolution is at least as old as
Aristotle, who, in equal opposition to the eternal
"flux" of Herakleitos and the eternal "rest" of the
Eleatics, taught that transition from that which is
not yet to that which is, or development, was the
only reality. Prior to the spotless and immortal
Darwin, however, whose epoch-making book was
the foundation of modern scientific evolutionism,
the most influential form of the development theory
was that of Idealistic Evolution, — the evolution
of the universe as a phenomenal representation, not
as a noumenal fact.

The trouble with Idealism is, and always has
been, that it never dares to be strictly logical —
never dares to march straight, from its premise in
the Cartesian "*Je pense, donc je suis,*" to its only
logical conclusion in solipsism. Even Schopenhauer,
who starts off so boldly in his "The world is my
representation,"[1] shows his timid inconsistency in
the very same sentence, admits the existence of
other thinkers, infers a world from which escape
is the one thing needful, and thus lands us in an
intellectual pessimistic quagmire to which his halt-
ing Idealism has been the guiding will-o'-the-wisp.
A valiantly logical Idealism might, perhaps, be irrefutable, but it would certainly be absurd; for

[1] "'Die Welt ist meine Vorstellung'—ist, gleich den Axiomen
Euklids, ein Satz, den Jeder als wahr erkennen muss, sobald er
ihn versteht; wenn gleich nicht ein solcher, den Jeder versteht,
sobald er ihn hört." (*Werke*, III. 4.)

a dialogue between two solipsists, each conceiving the other to be merely a "thing in his own dream," would be the very climax of the comical. Idealism ought to be a monologue; it has no rational right to be a dialogue at all, unless after the fashion of Dr. Johnson, who, coming to breakfast one morning in an ill-humor because he had dreamed overnight that he was beaten in argument, could not be consoled until he remembered that he had been both his antagonist and himself in one, and had therefore only beaten himself after all. Let us imagine ourselves as overhearing what we will style the —

§ 67. *Soliloquy of the "Consistent Idealist."*

"I think, therefore I am. I cannot doubt this original and necessary starting-point in my philosophizing; for, if I doubt, I think, and, if I think, I am. By doubt itself I am brought back to this very starting-point, since from it my doubt itself must start. My philosophy must evidently begin with this immediate knowledge of myself as thinking and existing, knowing and being, in one indivisible reality; my first fact must be that of my own consciousness as immediately manifesting itself to itself in a real identity of knowing and being. I can in no way account for this first fact, any more than I can doubt it; it is only a given fact, for which no reason is assignable; it is a fact which is indubitable simply because it is immediately self-evident; any reasoning I could devise would beg

the question, since it would presuppose the very
fact it was seeking to explain; I cannot prove that I
exist reasoning — I can only *reason*. Consequently,
I can get neither behind nor below this first fact
as my rational foundation in philosophizing.

"But is this 'I think, therefore I am,' the whole
of my first fact? Perhaps I have left something
out which is really part of it. If so, my philosophy
will be all wrong. Let me scrutinize this first fact
more keenly.

"It is self-evident that I cannot think without
thinking something. My thinking is an activity,
and must act on an object. What is this something,
this necessary object of my thinking? Whenever I
think, I discover that I always think a world, now
in part and now in whole; 'I think a world,' then,
is the general formula of all my thinking. It is not
enough to say, 'I think, therefore I am;' I must say,
'I think a world, therefore I am.' But why must
I not say, 'I think a world, therefore I am and the
world is'? It seems as if I must, since both I and
the world are in my consciousness.

"But no. I do not find that I am conscious of
any world; I am only conscious of myself thinking
the world; as distinct from myself, the world is not
in my consciousness, after all. My thought of the
world is only my own thought, — only my own
representation; I do not find in it anything but
itself, anything but thought, anything but representation; I do not find that I know any object

of my thought as distinct from my thought itself. It is evidently nothing but my mere representation. Indeed, I do not know why I should call it a 'representation' at all; it is certainly not a representation of any reality outside of myself that I can immediately know. In the pure content of my thought, nothing is presented, and therefore nothing can be represented, except myself alone. If it is a representation at all, then, it must be a representation of myself — though I do not recognize the likeness! The world, which seems to be represented as an object distinct from myself, can be in reality nothing other than myself, after all. Very well, then; when I say 'representation,' I will keep it clearly in my mind that this expression must mean, in all possible cases, nothing whatever except a 'representation of myself' — never a representation of anything other than myself. My thought always gives the *Me*, never the *Other*.

"But somehow, notwithstanding my irrefutable reasoning, I find myself in difficulty. This representation of the world I have just been demonstrating to be only a representation of myself. But, for all that, it quite obstinately refuses to put on the appearance of a representation of myself as I otherwise know myself. It quite obstinately refuses to give me either a front-view, or a side-view, or a back-view, or an over-view, or an under-view, of myself; it most obstinately persists in giving me a view of myself which I cannot get otherwise at all, — in giving me a view which, but for my good reasoning, I should certainly believe to be the likeness of an Other.   Well, then, if I cannot really *know* it as an Other, may I not at least *infer* it as an Other, and thus get out of my awkward difficulty?   May I not thus, without committing suicide as a reasoner, logically attribute to it a semi-known, but real, existence external to myself?   May I not infer that this external existence in some inscrutable way affects my own existence, determines my representations, and makes me think as I do?   What a relief that would be!   What an easy way out of my difficulty, and what an easy explanation of this puzzling and very annoying obstinacy in my representation of the world!

"Let me be cautious, however, and not destroy the foundation of my whole philosophy.   This 'inference' of mine is, after all, only another of my representations; and so, of course, is the thing inferred.   I am not conscious of the thing inferred; I am only conscious of myself as inferring it.   My thought of the thing inferred is nothing whatever but merely my own thought; it contains nothing but my own thought; it does not contain the thing inferred; it is nothing but myself once more.   In short, the inference refuses to infer!   It is just as obstinate as the representation of the world.   The representation refuses to represent, and the inference refuses to infer.   Neither of them has the least pity on my perplexity.   If I could say now that the *thing inferred* is an Other, I could just as well have said at the outset that the *world represented* is an Other. Either would be the complete sky-high explosion of my philosophy. I must stick to my principle and deny that Other, whatever siren song it may sing. My whole philosophy is at stake, and it calls upon me to be heroically logical at this critical point.

"So be it then! *Whatever I think, or represent, or infer, or imagine, or believe, contains Me, and no Other, as both subject and object:* that is Idealism, and anything else is nothing but its phantom and its sham. The philosophy which admits into my thought any Other whatever is essentially Realism, and not Idealism. As an Idealist, I must confess that whatever I infer is myself in disguise. I cannot break out of the closed magic circle of my own mere thinking. My representations of the world, of the inference, of the thing inferred, are only representations of *Myself.* I must, then, be something more than I imagined! My consciousness cannot be the whole of me; there must be in me an unconsciousness too, out of which these obstinate representations of the Other are involuntarily produced. I do not voluntarily or consciously produce them, yet they obstinately persist in appearing. Very well: they must emerge out of some unsuspected depth of my own being, in obedience to some force or law of my own being which I do not consciously comprehend. It is clear that I am immensely greater and grander than I at first suspected or imagined!

"But some of my representations wear the guise of intelligent beings like myself; they seem to talk with me, deal with me, act upon me. Whatever they do or say, however, is nothing but my representation; and my representation, if I am an Idealist at all, can be nothing but myself. The conclusion is irresistible; these other intelligent beings I call mankind exist only in my thought; they are unreal except as I give them reality; they exist in my perception, and are annihilated in my non-perception; they have absolutely no being but in my representation; their *esse* is *percipi*. I am their sole Creator, as I am the sole Creator of the world itself.

"I am equally their Destroyer. I represent them as dying, and therefore I am the Creator of Death. But I cannot represent myself as dead; that would be the representation of something which is not *Me*, but an *Other;* I cannot create that; therefore, I cannot die. Death to me would be the non-representation of myself; all my representations are of myself; I cannot represent my own death or non-being; I can only represent myself as living, and not dead; and all that I cannot represent is to me absolute nothing. Therefore, all the representations I call men die when I cease to represent them; but I shall never die, because I cannot cease to represent myself. I am the Eternal.

"Infinite Space and Time are no less my representations, and therefore I create them. Infinite Being itself is my representation, and I create that too. As a unity of conscious and unconscious powers, I create the world, mankind, Space, Time, God himself; for all these are nothing but my representations, and all my representations are nothing but myself. I am all that I create. Hence I myself am the All: I am the Infinite, I am the Absolute, I am the Eternal, I am the only true God!

"Having arrived at this satisfactory and only logical conclusion from my Idealistic principle, and having triumphantly and successfully swallowed the universe, I will now take a nap."

§ 68. Such would be the soliloquy of a "consistent Idealist," and it is modestly suggested as an object-lesson in logic to inconsistent Idealists. But, alas, the "consistent Idealist" is himself an absolutely ideal being: he is nothing but "my representation," and I have never met him either in real literature or in real life. The real Idealist I meet is always inconsistent—always dilutes his Idealism with a dash of Realism; he boldly applies it to the world of matter, but never dares to apply it unflinchingly to his fellow-men in general, or to his interlocutor in particular; he boldly applies it to Space and Time, but seldom or never to God. Now, as has been said above (§ 66), "a valiantly logical Idealism might, perhaps, be irrefutable, but it would certainly be absurd;" and few would deny that the above soliloquy ends in absurdity. But any philosophy becomes absurd and unworthy of intellectual respect, when it wilfully shirks the logic of its fundamental principles and makes arbitrary exceptions to them; and real Idealism, in a different way, is just as absurd as solipsism. The only "consistent Idealist" is the solipsist himself; there is no other; and our overheard soliloquy shows what, if he could be found, he would say. But the solipsist himself, if he ventured to say it to "an Other" than himself, would thereby concede that "Other's" existence, and therefore forfeit the laurels due to his courage and consistency so long as he only soliloquized. In the determined silence of the solipsist lies the only irrefutability of Idealism. Real Idealism is already refuted, if the detection of self-contradiction is refutation; but who can refute a man who refuses all dialogue?

§ 69. The whole plausibility of Idealism lies in its assumption of its unscientific "first fact:" the Idealist begins with his individual consciousness alone as the only certain or indubitable datum, while science begins with universal human consciousness and the universe it has discovered. Descartes, who unwittingly launched modern philosophy upon its Idealistic voyage by his " I think, therefore I am," was himself a Conceptualist, a product of the Extreme Nominalism which was championed by Roscellinus and his companions hundreds of years before;[1] and Kant and his successors, men of mighty

[1] "De même, le nombre que nous considérons en général, sans faire réflexion sur aucune chose créée, n'est point hors de notre pensée, non plus que toutes ces autres idées générales que dans l'école on comprend sous le nom d'universaux." (Descartes, *Œuvres*,

genius, could but follow the general direction thus imparted to all modern philosophic thought. Hegel, the greatest of the post-Kantian Idealists, says: "Thought, by its own free act, seizes a standpoint where it exists for itself, and generates its own object;"[1] and again: "This ideality of the finite is the chief maxim of philosophy; and for that reason every true philosophy is Idealism."[2] This is the absolute sacrifice of the objective factor in human experience. Hegel sublimely disregards the distinction between Finite Thought and Infinite Thought: the latter, indeed, *creates*, while the former *finds*, its object. And, since human philosophy is only finite, it follows that *no* true philosophy is Idealism, except the Infinite Philosophy or Self-thinking of God.

But all modern scientific thought has, in spite of Bacon's seeming hostility to Aristotle's influence, substantially held the Aristotelian ground. Cutting, like Alexander, the Gordian knot, it has, like Alexander, conquered a world: it construes experience as inclusive of object and subject both, and refuses to construe it, as Idealism does, as inclusive of the subject alone. Here is the exact point of divergence

III. 99, ed. Cousin.) "On compte ordinairement cinq universaux, à savoir, le genre, l'espèce, la différence, le propre, et l'accident." (Ibid., III. 101.) This principle of Conceptualism denies by necessary implication the *objectivity of relations*, and therefore of all the *objective relational systems* discovered by the scientific method. Logic and history alike show that every possible philosophy is built either on the subjectivity or the objectivity of relations, and the world will yet find out this fact.

[1] *Werke*, VI. 25.   [2] *Werke*, VI. 189.

between the Idealistic and Scientific Hypotheses, for
hypotheses they equally are; the truth of perception
cannot be logically proved. But if the wonderful
increase of human knowledge by the use of the
scientific method be not verification of the original
scientific hypothesis, then there is no such thing as
verification, and all human knowledge is a melancholy lie.

§ 70. To-day, then, if science can establish anything, it has established the principle of Realistic
Evolution, to the complete overthrow of the principle of Idealistic Evolution; and scientific realism
treats the evolution of the universe, not as a merely
phenomenal fact, but as a fact which is at once both
phenomenal and noumenal. Let us take up once
more the thread of our argument at this point, and
go on to determine the conception of Universal
Realistic Evolution in a way that shall satisfy the
demands of science and philosophy alike.

Two possible views of Universal Realistic Evolution, to which all others are logically reducible,
present themselves for consideration: namely, the
Mechanical and the *Organic.* The triumph of the
profounder view at last will be the determination
of the concept of the immanent relational constitution of the universe *per se* as either a *Machine* or an
Organism. The one theory conceives the universe
as a machine, and seeks to explain it on simply
mechanical principles; the other theory conceives
the universe as an organism, and seeks to explain

it on organic principles. The eternal warfare of
ideas by which all intellectual progress is effected
centres to-day in the struggle between these two
opposing theories or tendencies. What has noumenism, the genuine philosophy of science, to say
respecting them ?

§ 71. First of all, this: that, just as noumenism
itself affirms all that phenomenism affirms (the
reality of phenomena), and 'affirms also what phenomenism denies (the reality of noumena), so the
organic theory of evolution affirms all that the mechanical theory affirms (the facts and principles of
mechanism), and affirms also what the mechanical
theory denies (the facts and principles of teleology
and the absolute failure of mechanism to explain the
universe without them). In other words, the mechanical theory covers only a part of the facts, while
the organic theory covers them all.

§ 72. Both theories accept the fact of an objective
and intelligible relational constitution of Nature,
totally independent of human representation for its
existence; both, therefore, so far as they are self-
consistent and worthy of philosophical recognition,
accept without question the principles of noumenism,
and are equally bound to accept their logical results.
Both, furthermore, accept the fact that this objective
relational constitution of Nature is a veritable system, in which all the parts are so closely correlated
as to constitute a rational whole. But here their
divergence begins. The mechanical theory denies

that the system of Nature is a perfect system, and
heaps up proofs of its imperfection in the existence
of evil; the organic theory affirms that the system
of Nature is perfect within the limits of possibility,
and claims that the existence of evil results from the
absolute conditions and logical necessities of finite
existence as such — does not, therefore, prove any
avoidable or real imperfection in the system of Nature. Furthermore, thé mechanical theory takes, as
type of the actual system of Nature, the machine;
while the organic theory takes the organism. It is
evident enough that the mechanical theory conceives
the real and rational unity of Nature in a far lower
and cruder form than the organic theory; for the
organism is a machine *plus* a great deal more, and
yields a concept of far higher closeness, complexity,
and comprehensiveness of internal relationship, and
therefore of far superior richness of content.

§ 73. No little light is thrown upon the nature
and relative value of these two theories, viewed as
mere hypotheses, by a critical analysis of the two
fundamental concepts on which they are based, and
to which, despite all special pleading, they must be
ultimately reduced. The fact that *no machine either
lives or grows,* while *every organism both lives and
grows,* shows at once the formidably embarrassing,
nay, the overwhelmingly crushing, difficulty under
which the mechanical theory labors, in trying to
work out an intelligible and complete concept of
evolution as the life and growth of the universe; for

the very concept of evolution as life and growth is essentially organic, has been derived only from the organism, and is, in truth, utterly incompatible with that rigid exclusion of all but mechanical principles which it is the specific purpose of the mechanical theory to establish. All is easy enough, so long as this theory deals with the purely physical or mechanical facts of Nature; but, the moment it approaches the domain of biology, its difficulties begin, and soon grow so formidable, in the domain of psychology, sociology, and ethics, that the theory itself, even in the hands of really able champions, obtrusively and hopelessly breaks down. The fact is that the extension of the idea of evolution to the inorganic world, and to the system of Nature as a whole, betrays unmistakably the inward (though perhaps unconscious) pressure of the organic idea in the scientific mind; and hence nothing but intellectual confusion has resulted, or possibly could result, from the attempt to conceive evolution as exclusively mechanical. No wonder, then, that it is impossible to find an ostensibly mechanical interpretation of Nature which does not, the moment it approaches biology, yield to the temptation of surreptitiously introducing organic elements into its professedly mechanical system, and thereby demonstrate its inability to remain faithful to the facts without surrendering its own fundamental principle. This result is simply inevitable from the nature of the case. It is the fault of the facts, which persist

in *not* being purely mechanical. The concept of
evolution, applied to the universe as a whole, is
necessarily the concept of it as a living and growing
whole; it must include all facts, physical no less
than biological and psychical, under this concept
of life and growth ; and the concept of life and
growth, in which alone the Many and the One are
absolutely reconcilable, is essentially and necessarily
that of the organism. The search for the One in
the Many and the Many in the One has been from
antiquity the essential task of philosophy; and I
feel perfectly safe in asserting that no idea ever has
been or ever will be found which shall absolutely
reconcile the Many and the One except the idea of
the organism itself. Most certainly the idea of the
machine, no matter how elaborated or how expanded,
can never be made by any degree of ingenuity or
acumen to cover those facts which are of supreme
interest to philosophy, and which are just as deeply
inwrought into the warp and woof of the universe as
the plainest facts of chemistry, physics, or mechanics.
When Alexander von Humboldt, in his *Kosmos*,
called the universe "a living whole," he showed a
flash of philosophic insight in a purely scientific man
which puts to shame the obtuseness of more than
one reputed philosopher.[1] Science itself, as science,

---

[1] " Die Natur ist für die denkende Betrachtung Einheit in Viel-
heit, Verbindung des Mannigfaltiges in Form und Mischung, Inbegriff der Naturdinge und Naturkräfte als ein lebendiges Ganze."
(*Kosmos*, I. 5, 6, ed. 1845.) (Compare Hegel, *Werke*, VII. 38:
" Die Natur ist an sich ein lebendiges Ganzes.")

is now brought face to face, by the established fact
of universal evolution, with a question, also of fact,
which yet admits only of a philosophical solution :
namely, *is this universe a machine or an organism ?*

§ 74. The old distinction of Nature as " organic
and inorganic," conceived as two departments of
existence which can be really and exactly demar-
cated, has become utterly discredited and outgrown,
as a distinction which is intrinsically misleading,
artificial, and false in itself. It is no longer possible
to point out where the line is to be run between
animal and plant, or between living and non-living
matter. The old fence is down, and no man has
skill enough to rebuild it. But what follows ? A
most momentous consequence : namely, that *the
universe is either wholly organic or wholly inorganic.*
Which shall it be ? " Inorganic !" says the mechanical theory. " Organic !" says the organic theory. The one would level all things down to the
grade of the machine; the other would level all
things up to the grade of the organism. The one
would explain the organism itself as merely a more
complicated machine; the other would explain the
machine itself as merely a lower and less developed
form of the organism — as an artificial organism
created by the natural organism. The issue is a
vital one, and it is hotly fought to-day all over the
civilized world, wherever thought is not swamped in
mere brute existence. There is no possibility, however, of finally settling this profound and vital issue

by appealing to any discoveries made by the empirical use of the scientific method; nothing but the settlement of what was called at the outset (§ 8) the "previous question" of phenomenism and noumenism, — nothing but the philosophizing of the scientific method itself, — will ever lead to a permanent settlement of the question whether the universe must be viewed as wholly organic or wholly inorganic. And on the right settlement of this question at last depend all the highest ideal hopes, all the highest moral interests, all the highest religious aspirations of mankind.[1]

§ 75. But the close comparison or analysis of the two concepts of the machine and the organism still remains to be made.

If the results of our inquiry into the nature of intelligence, in the preceding chapter, are valid, every relational system is a product of the creative understanding, and every product of the creative understanding is essentially a means or an end. Now both the machine and the organism are relational systems; that is agreed. Both of these relational systems are teleologically constituted, as means or ends; but that is disputed, since the modern mechanical theory stoutly denies all teleology, even in the structure of the organism. Analysis,

[1] " But this I do say, and would wish all men to know and lay to heart, that he who discerns nothing but Mechanism in the Universe has in the fatalest way missed the secret of the Universe altogether." (Carlyle, *On Heroes, Hero-Worship, and the Heroic in History*, p. 160, New York, 1872.)

however, shows that teleology is just as deeply
wrought into the system of the machine itself as
into that of the organism, and is the only possible
explanation of either. Consequently, even if the
mechanical theory is correct in maintaining that
every organism is a mere machine, its contention is
tantamount to an unconscious confession that every
organism is teleologically constituted, — tantamount,
therefore, to an unconscious, yet absolute, surrender
of its own fundamental idea. This criticism is a
fatal one, but I will waive it for the present.

§ 76. The machine is a system in which the parts
are so related that the whole, as a cause, is adapted
to the accomplishment of purely external ends, as an
effect; and the pure externality of these ends proves
an external mechanist, in whose mind the ends exist,
and by whose hand the machine has been made
what it is, in order to accomplish those ends. It
does not effect those ends by itself, but only as used
by the mechanist to effect them; it does not form
itself, repair itself, or reproduce itself; it exists only
by another, and for another; it is purely artificial —
the work of art for purposes of art. Such is every
machine certainly known by man to be a machine,
from a simple nail to a vast railroad system: the
only concept of it drawn from human experience is
that of *a means adapted to external ends.* So overwhelmingly strong is the induction based on this
experience that a mere bit of flint, rudely resembling
the head of an axe or arrow, and dug out of a deep

excavation together with innumerable other stones, carries conviction to every civilized mind that this crude machine is demonstration of the existence of prehistoric savages, by whom it was fashioned, as a means, for chopping or slaying, as an end external to itself. The machine, however, is itself an organism of lower grade, — an artificial extension, as it were, of the living organism; as, for instance, the carpenter's tool is an extension of the human hand, creatively conceived by the human mind and creatively wrought by the human hand itself. But the tool does not explain itself, much less the hand; while the hand and the mind do explain the tool. In short, the machine is an irresistible proof of the mechanist, and is both inexplicable and inconceivable without him; while, on the other hand, the machine and the mechanist together constitute in truth only a larger organism which has, by art, extended the boundaries of its own existence. Consequently, the machine, though in no sense an organism in itself alone, is yet, in a very true sense, an organism of a lower grade, inasmuch as the true or living organism has annexed it to itself as a conquered province of the not-itself, and so far given it a temporary and imperfect, though strictly subordinate, organic being. Hence it is clear that the machine cannot be made to explain the organism, but, on the contrary, can itself be explained by the organism alone; it does not exist as an end in itself, but solely as a means to an end external to itself;

and it becomes an organism of a lower or non-living grade, only when used by a true or living organism as an artificial extension of itself in the execution of its own organic ends.

§ 77. Now, when the mechanical theory applies this concept of the machine to the philosophical explanation of the universe, it must of course conform to the requirements of philosophy; it must not logically violate the essential nature of the concept it employs. Consequently, as a mere machine, the universe should be conceived by the mechanical theory as simply a means to an end, and as implying, like every other machine, its own external mechanist. The only way to realize this concept, logically or philosophically, is to complete it by conceiving God as the external mechanist or creator of the universe, and the "glory of God" as the end for which he has created it. Hence the mechanical theory in its only logical form is pure and absolute *Dualism;* and its Dualism is in the form of an old-fashioned, artificial, truly mechanical, and wholly outgrown type of theology. If, on the contrary, the mechanical theory, in order to deny teleology, discards Dualism and professes *Monism* (as, curiously enough, it does in all modern mechanical philosophies), it thereby reduces itself to the utterly unreasonable and unintelligible position of declaring the universe to be a *means,* yet a *means to no end !* For the machine is essentially nothing but a means to an external end, as has just been shown; and

there can be no end external to the universe. From
this conclusion there is no possible escape. The
mechanical theory, when logical, is either old-
fashioned supernaturalism, or else natural teleology
of the Paley type; and, if it presumes to stigmatize
its rival as " anthropomorphism," the retort is crushing that *it is the mechanical, not the organic, theory
which likens the universe to the machine — that is, to
the " work of men's hands."* It would be safe for
the mechanical theory not to indulge itself in that
particular sarcasm. As a professedly " Monistic
Philosophy of Evolution," this theory philosophically
destroys itself by adopting the machine as its concept of the universe; for the concept of the machine,
if applied logically to the universe as a whole, is the
necessary denial of Monism. And, finally, since both
the machine and the organism necessarily presuppose
teleology and are equally inconceivable without it, the
mechanical theory of evolution utterly breaks down :
its denial of teleology is its suicide as a philosophy.

§ 78. Such is the concept of the machine, and
such is the philosophical result of the attempt to
apply it to the explanation of the universe. What,
then, is the concept of the organism, and what will
be the philosophical result of the attempt to apply
that to the explanation of the universe ?. It would
be impracticable, within the necessary limits of the
plan of this book, to go fully into this subject; but
enough can be said in a reasonable compass to serve
our present purpose.

The organism is a system in which the parts are so related that the whole, as a cause, is adapted to the accomplishment of either external ends, internal ends, or both. The non-living artificial organism (the machine) has only external ends; the living natural organism (the plant, animal, man) has both external and internal ends; the living cosmical organism (the universe) has only internal ends. The machine has already been explained: it is only a means to an end, and the end is not its own, but only that of the natural organism which has created it, and which gives it, by using it, the only life which it can be conceived to have. But the natural organism is created by the cosmical organism, *first*, as an end in itself, and, *secondly*, as a means to an end which is not its own, but that of the cosmical organism which has created it. As an end in itself, the natural organism lives simply to " full-fill " its own life; as a means to an end which is not its own, but that of the cosmical organism, it is simply a machine with reference to the latter, in precisely the same sense in which the pure machine is a mere means to an end of the natural organism itself. The cosmical organism eternally creates itself simply to " full-fill " its own life; every relational system which it thus creates within itself, whether mechanical or organic, is from this point of view merely a means to this supreme end of all Being, and, therefore, merely a machine; but, in freely or creatively " full-filling " its own life, so far the natural organism freely or

creatively helps to "full-fill" the universal life. The
pure machine, then, or artificial organism, is a pure
means to an end not in itself; the natural organism
is both an end in itself ("full-filling" its own life)
and a means to an end which is not in itself (helping
to "full-fill" the cosmical life); while the cosmical
organism is at once an absolute end in itself and an
absolute means to this end in itself.

The natural organism must, therefore, be conceived as having both an *Indwelling or Immanent
End* and an *Outgoing or Exient End.* Its immanent
end (formative and reparative) is the "full-filling"
of its own life, renders it at once both cause and
effect of itself, and constitutes the principle of egoism, legitimate selfishness, or self-preservation and
self-development. Its exient end (reproductive and
co-operative) is the helping to "full-fill" the universal divine life, and constitutes the principle of
altruism, legitimate unselfishness, or self-sacrifice
and self-devotion. These two principles show themselves in active exercise in all organisms which have
reached even a low position in the scale of being.
In man, particularly, the immanent end shows itself,
in the individual, by the pursuit of happiness, of
knowledge, of moral and religious culture in general,
no less than of lower personal aims, — in society, by
the foundation and fostering of institutions of all
sorts for the preservation and spread and progress
of civilization, and so forth; while the exient end
shows itself in the reproductive and philoprogenitive

instincts, in patriotism, in philanthropy, in devotion
to the discovery of truth, and so forth, but above all
in the supreme activity of love, veneration, and self-
consecration. The way in which this exient end in
the natural organism is used by the cosmical organism, in the furtherance of cosmical ends irrespective
of the individual, is manifested with especial clearness in making it subservient to the preservation of
the species and the perpetuation of life in general.
The reproductive system is no benefit, but rather a
detriment, to the individual as an individual; it is
a diversion of individual vitality to the service of the
general good; it is the subordination of the individual to the self-preservation and self-development
of a higher individual, or relational system, in the
species or kind. In this is shown how the natural
organism is used by the cosmical organism as a mere
machine — a mere means for the realization of ends
in which the individual has no individual interest,
and in which he can sympathize only through a
high religious sympathy in the "full-filling" of the
cosmical life itself, the general well-being of the
universe as a whole.

§ 79. Thus the organic conception extends itself
from the atom or molecule, the simplest discoverable
machine, up to the universe of Being as a whole,
the supreme cosmical organism; and the idea of the
organism, as that in which alone the Many and the
One are reconcilable, covers and includes all the facts
which science has discovered or may yet discover.

13

While, therefore, the mechanical theory proves itself utterly unable to explain even its own fundamental concept, that of the machine, and much less that of the organism, without calling in the assistance of the teleological idea which it claims to reject, the organic theory finds in this very idea the "open sesame" of philosophy — the rational and real unity, not only of all organic facts, but of all facts whatever; and it shows that teleology, so far from being overthrown by the fact of Evolution or the theory of Darwinism, is the only principle which renders either Evolution or Darwinism philosophically intelligible. It is, in truth, the only principle which lights up the universe from within, and renders it luminous and transparent, so to speak, from centre to circumference.

§ 80. If any further proof is wanted of the absolute necessity of the principle of teleology in science itself, it is forthcoming in the fact that no mechanical theory of evolution has yet appeared, so far as my knowledge goes, which does not deny itself, beg the question, and surrender the whole point at issue, by consciously or unconsciously, overtly or covertly, introducing of itself the teleological principle, the moment it approaches the province of biology. I will only mention Herbert Spencer and Ernst Haeckel, the two ablest defenders of the mechanical philosophy.

§ 81. What is Spencer's definition of life? "Life," he says, on the one hand, "is definable as the continuous adjustment of internal relations to external relations."[1] On the other hand, he distinctly and unequivocally rejects all teleology, as a principle of scientific explanation of the universe.[2] But "adjustment" is a concept which is absolutely identical with the teleological principle. "Adjustment of internal relations to external relations" can only be a *change* in internal relations, made, *as a means*, to effect a *correspondence* with external relations, *as an end*. The change is the means, the correspondence is the end; and that is teleology in undiluted strength. The very essence of life, then, consists in teleological activity; and this teleological activity *must* be conceived, according to Spencer, as that of the very "Unknowable Power" which, still according to Spencer, *cannot* be conceived as acting teleologically. No machine ever "adjusts" itself to anything not foreseen and provided for in the mechanism itself by the mind which has created it; it simply suffers damage or destruction. "Adjustment" has no conceivable meaning but the adaptation of means to ends; and, if the power of "adjustment" is admitted to be so wrought into the organic structure as not to be referable to the organism's own consciousness, that is an admission that an external mind has wrought it there, — that the organism is *not* a mere machine, — that Nature works teleologi-

---

[1] *First Principles*, p. 84, 4th Ed. So *Principles of Psychology*, I. 293.

[2] *Principles of Biology*, I. 340.

cally, and not mechanically, in providing beforehand, in the very organism itself, for the exigencies of organic life. Thus Spencer has written down the absolute and irretrievable failure of his whole philosophy, as a mechanical theory of evolution, in that one word "adjustment."

§ 82. Haeckel, likewise, the bolder and more sequent thinker, does the same thing just as conspicuously in his own philosophy. On the one hand, he says : "Moreover, we shall have good reason to hope that at some future time we shall learn to explain the first causes at which Darwin has arrived, namely, the properties of Adaptation and Inheritance ; and that we shall succeed in discovering in the composition of albuminous matter certain molecular relations as the remoter, simpler causes of these phenomena. There is indeed no prospect of this in the immediate future, and we content ourselves for the present with the tracing back of organic phenomena to two mysterious properties," etc.[1] "*Inheritance* is the *centripetal* or *internal formative tendency* which strives

---

[1] *History of Creation*, I. 32, Amer. ed. The italics above are as there printed. Prof. Enrico Caporali, in his brilliant series of articles on "*La Formola Pitagorica della Cosmica Evoluzione*," still publishing in *La Nuova Scienza* (which is the organ of the most hopeful intellectual movement, in the direction of a truly scientific and yet truly religious philosophy, which appears within the philosophical horizon of the present), adds to Haeckel's two causes the missing third — "Heredity, Adaptation, and Selection" — by which "Selection" Caporali means, not the action of mere mechanical causes, but the teleological activity of the "*Unità Madre*," or Nature as Prolific Unity. (*La Nuova Scienza*, I. 75 : "— *tre processi cosmici, Eredità, Adattamento, e Cernita.*")

to keep the organic form in its species, to form the descendants like the parents, and always to produce identical things from generation to generation. *Adaptation*, on the other hand, which counteracts Inheritance, is the *centrifugal* or *external formative tendency*, which constantly strives to change the organic forms through the influence of the varying agencies of the outer world, to create new forms out of those existing, and entirely to destroy the constancy or permanency of species. Accordingly as Inheritance or Adaptation predominates in the struggle, the specific form either remains constant or changes into a new species. The degree of constancy of form in the different species of animals or plants, which obtains at any moment, is simply the necessary result of the momentary predominance which either of these two formative powers (or physiological activities) has acquired over the other." [1] On the other hand, Haeckel says with reference to the Theory of Descent: "As soon, in fact, as, according to this theory, we acknowledge the exclusive activity of physico-chemical causes in living (organic) bodies, as well as in so-called inanimate (inorganic) nature, we concede exclusive dominion to that view of the universe which we may designate as the mechanical, and which is opposed to the teleological conception." [2] And, in his *General Morphology of*

---

[1] *History of Creation*, I. 253, 254.
[2] Ibid., I. 17. So, also, pp. 69, 100, 167, 176, 262, 337 — in fact, *passim*.

Organisms, Haeckel devotes a whole chapter to what he calls the "Purposelessness, or Dysteleology," of Nature.

Here, then, we have the same contradiction which we have just found in Spencer. "Inheritance," the first of the two formative causes which he believes to explain the whole fact of organic evolution, is nothing but the *means* by which Nature reproduces the organic structure of the species, which is her *end;* and, strangely enough, Haeckel himself admits this in the very passage above quoted, when he defines Inheritance as the natural "tendency which *strives to keep* the organic form in its species"! For, plainly enough, the *striving* is the *means,* and the *keeping* is the *end.* Could anything be more evident than the fact that Haeckel unconsciously conceives "Inheritance" itself as a natural teleological activity? So, also, "Adaptation," the other formative cause, is the *means* by which Nature secures the gradual appearance of new species, which also is her *end;* and here again Haeckel, with amusing unconsciousness, himself describes it as the natural "tendency which *strives to change* the organic forms" — the striving being the means and the changing the end! Machines do not propagate their kind, do not inherit ancestral forms, do not adapt themselves to circumstances. Strike out the teleological significance from these two words, "Inheritance" and "Adaptation," and they lapse into absolute meaninglessness. By using them, or rather by not understanding them (for the *petitio principii* is in the essential conceptions of them, as their above given definitions prove), Haeckel shows once more the utter impossibility of explaining biology without the help of teleology, and, like Spencer, disproves his own mechanical theory. Not by any chance slips or careless expressions, but by the most fundamental concepts of their systems, these two foremost champions of the mechanical theory tear down with the one hand what they build up with the other, and demonstrate the impossibility of constructing a mechanical philosophy of evolution which shall not fundamentally assume the very teleology it professes to reject.

§ 83. The truth is, neither Spencer nor Haeckel ever yet clearly conceived any form of teleology except the old-fashioned, dualistic, supernatural, and really mechanical teleology of the Calvinistic or of the Paley school; neither of them has the faintest conception of the new, monistic, strictly natural, and purely organic teleology of scientific philosophy. Their systems, therefore, are out of date already; they are not abreast of the age. Haeckel shows this incontrovertibly in the following passage, and it is no less evident in Spencer: "The artificial discord between mind and body, between force and matter, which was maintained by the erroneous *dualistic and teleological* philosophy of past times has been disposed of by the advances of natural science, and especially by the theory of development, and can no longer exist in face of the prevailing *mechanical and monistic* philosophy of our day."[1] (The italics are mine.) The "dualistic and teleological" philosophy of Paley belongs indeed to the past; the "mechanical and monistic" philosophy of Spencer and Haeckel belongs to the present, but is rapidly moving into the past; the *teleological and monistic* philosophy of the scientific method and the organic theory of evolution belong to the future, and will soon be here. But, apparently, neither Haeckel nor Spencer ever dreamed of that. The true relations of Dualism, Monism, and Teleology have been alluded to earlier in this chapter (§ 77), and it is sufficient to refer here to that former statement. The organic theory of evolution, which is monistic and teleological at the same time, is the only form of Monism which can logically exist at all. The teleology which it presents is endocosmic, not exocosmic, — immanent in the universe as its omnipresent thought and life, not external to it as that of a Mechanical Creator, working in material alien to and other than himself. Inasmuch as every machine logically implies a machinist, mechanist, or mechanic, the mechanical theory of evolution obsti-

[1] Ibid., II. 361. The only theism Haeckel can conceive is "the unscientific idea of a creator existing out of matter (*die unwissenschaftliche Vorstellung von einem ausserhalb der Materie stehenden und dieselbe umbildenden Schöpfer*)." (Ibid., I. 10.) Spencer shows scarcely more insight in his very shallow treatment of "the atheistic, pantheistic, and theistic hypotheses" in his *First Principles*, pp. 30–36. It only takes six pages, in his opinion, to exhaust that subject!

nately implies and requires Dualism, — a Machine Universe here, a divine Master Mechanic there; and the arbitrary denial of teleology, instead of making it Monism, unmakes it as philosophy altogether. The only Monism which is logically possible is teleological through and through; and Monistic Teleology, the Organic Theory of Evolution, is the heir of the future.

CHAPTER VI.

THE GOD OF SCIENCE.

§ 84. THE immanent relational constitution of the universe *per se* is, then, not that of an infinite machine, which is a self-destructive concept, but that of an infinite self-created and self-evolving organism, which is the only concept capable of effecting an absolute reconciliation of the Many and the One. The immanent life-principle of this cosmical organism is endocosmic and monistic teleology, the omnipresent and eternal teleological activity of the infinite creative understanding or Infinite Self-conscious Intellect; for the free creation of ends and means (relational systems both subjective and objective) has been shown to be at once the essential *Method of all Being* and the essential *Method of all Thought,* and therefore, through this unity of method, the absolute *Ground of the Identity of Being and Thought* (§ 46). The absolute end of Being-in-itself, therefore, is the absolute "full-filling" of Thought-in-itself, — that is, creation of the Real out of the Ideal; and the absolute realization of this end is the *Eternal Teleological Process of the Self-Evolution of Nature in Space and Time,* — in a word, the *Infinite Creative Life of God.*

§ 85. This is the meaning of the principle that the universe is an organism, and not a machine, — a principle which is the logically necessary result of the thorough philosophizing of the scientific method. It shows that the whole universe of Being is instinct with an infinitely intelligible and infinitely intelligent Energy, working actively, in every point of Space and every moment of Time, according to the intelligible principle of Ends and Means — ends that are cosmical in their reach and scope, means that are cosmical in their dignity and effectiveness. It shows that this " Infinite and Eternal Energy from which all things proceed" effectively reveals itself in Nature to the human understanding — is in no sense " Unknowable," but essentially knowable *per se*, and actually known to the precise extent to which science has discovered the immanent relational constitution, or organic idea, of Nature itself. It shows that Nature is not *a " manifestation" which does not manifest*, but rather the veritable, natural, and infinitely intelligible self-revelation of the noumenal in the phenomenal, of the absolute in the relative, of the infinite in the finite, of the eternal in the temporal. It shows that there is a fundamental spiritual identity between man and the universe in point of essential nature; that free creativeness is the supreme characteristic of intellect, whether finite or infinite, and effectuates itself in the actual creation of ends and means, as subjective or ideal relational systems; that free executiveness, or will, is the neces-

sary concomitant of intellect, whether finite or infinite, and effectuates itself in the realization of these
ends and means in Nature, as objective or real relational systems.   This is the profound truth underlying the crude conception of primitive religions
that "God created man in his own image."   Anthropomorphism and anthropopathism are no absolute
errors, but contain elements of truth which philosophy will earnestly seek to find, and reverently cher-
ish when found.   Infinite Wisdom and Infinite Will
are characteristic attributes of God which stand luminously revealed in the organic or teleological conception of the universe *per se.*   But teleology has
not yet yielded its richest fruit.

§ 86. In our study of the concept of the organism
(§ 78), we found that every organism has a twofold
end — the Indwelling or Immanent End and the
Outgoing or Exient End.   Nature provides for the
realization of this exient end of the finite organism,
so far as it is her own immanent end as the infinite
organism, by implanting in every finite organism of
the higher orders the love of its own kind, the desire
of offspring, the divine passion of maternal and pater-
nal affection, the deep and indestructible yearning to
repeat itself in that whose life is a renewal and con-
tinuation of its own — in that which is at once both
itself and not itself.   Now, if the universe of Being
is indeed an organism, nay, the one supreme and
infinite organism, this exient end, it would seem,
must needs be defeated ; for there is nothing beyond

itself to which it can go out, and it cannot reproduce itself in another infinite. But it is not, for all that, lost. This exient principle of the universal organism, this self-abnegating and sublimest and most exquisitely beautiful element in the organic idea, constitutes that attribute in the character of God which is the rational foundation of religious trust and hope and love. For, far from vanishing or expiring in impotency, it reappears with redoubled power; it diffuses itself internally throughout the infinite organism itself, as a deepened energy and enhancement of the immanent end; it manifests itself as that Natural Providence of Law and Love in One which is the support of every instructed, steadfast, and religious mind; it returns, so to speak, into the bosom of the universe as illimitable love of itself, — as ineffable satisfaction in its own fulness, beauty, and perfection, and as boundless tenderness for the spiritual offspring, veritable "children of God," who "live and move and have their being" in itself alone. What is this but infinite beatitude, infinite benignity, infinite love, — the All-Embracing Fatherhood-and-Motherhood of God?

§ 87. If such is the form in which the principle of exiency must show itself in the infinite organism, no less sublime and glorious is the form taken by the principle of immanency. "The absolute end of Being-in-itself is the absolute 'full-filling' of Thought-in-itself, — that is, the creation of the Real out of

the Ideal:" this we saw at the opening of this last
chapter. Now the Ideal appears as the subjective
relational system freely created by the creative
understanding; and the Real appears as the objective relational system effectuated in Nature by the
subordinate realizing activity of the executive will.
The blindly executive will, however, is nothing but
the objectively creative potency of the understanding itself: Thought is Force, and Force is Substance.
The absolute "full-filling" of Thought-in-itself, therefore, or the embodiment of the Ideal in the Real, is
the eternal self-legislation of Thought-in-itself into
Thought-in-Being — of the subjective relational system into the objective relational system of the Real
Universe. The ground of this realization can only
be the inherent and uncreated fitness of the Absolute Ideal to Be — that is, to become the Absolute
Real; and the perception of this absolute fitness of
the Ideal to become the Real — a profoundly ethical
perception — is the ground of the Eternal Creative
Act. Here, then, the infinite organism manifests
itself essentially as Moral Being — as a universe
whose absolute foundation is Moral Law, of such
absolutely self-inherent sanctity that the creative
understanding itself obeys it and the whole fabric
of creation embodies and enforces it; and the moral
nature of man, derived from this moral nature of the
universe itself, is the august revelation of the infinite purity, rectitude, and holiness of God. The
unspeakable sublimity of the moral nature of man

is, therefore, testimony to the immeasurably vaster sublimity of the moral nature of the universe itself; for, as the atom is to infinite Space, so is the grandest virtue of man to the infinite holiness of God.

§ 88. I do not forget the problem of evil: alas, who that is human can forget that? But neither do I forget that evil is simply the pressure of our own finitude, and that even the Infinite Love and Compassion could not relieve us of that without accomplishing the inherently impossible, to which omnipotence itself cannot extend; for, just as omniscience, rationally conceived, is the knowledge of all that is knowable, but not of the unknowable (the non-existent or nonsensical), so omnipotence, rationally conceived, is power to do all that is doable, but not to do the inherently undoable — that which involves self-contradiction or violates the necessary nature of things. Derivative being cannot, in the nature of things, either be or become infinite; and nothing short of infinitude could bring to us release from all evil. Evil is no end in itself; it cannot exist in the universe as an infinite whole, but only in the mutual relation of its parts, as the inevitable shadow-side of all finite reality. If it could be avoided, — if the finite real could possibly exist at all without the finitude which weighs upon it and is the source of all its woes, — then might we justly blame the universe for the evil that is simply inevitable. Is it not enough to lay this " spectre of

the mind" to know that, without this finitude, finite
being could not be; that finite being is better than
non-being; and that, between these two grim but
sole possibilities, Infinite Goodness and Love itself
would choose the former? If that is not precisely
optimism, neither is it pessimism; and it is theodicy
enough to satisfy at least one not easily satisfied
mind.

§ 89. Let us now review the general course of
thought which we have been pursuing in these
investigations, and gather together in a brief summary the large elements of that noumenal conception of the universe which naturally flows from
the philosophized scientific method.

1. Because the universe is in some small measure
actually known in human science, it must be in
itself both absolutely self-existent and infinitely intelligible; that is, it must be a noumenon because
it is a phenomenon.

2. Because it is infinitely intelligible, it must be
likewise infinitely intelligent.

3. Because it is at the same time both infinitely
intelligible and infinitely intelligent, it must be an
infinite subject-object or self-conscious intellect.

4. Because it is an infinitely intelligible object,
it must possess throughout an immanent relational
constitution.

5. Because it possesses an infinitely intelligible
relational constitution, it must be an absolutely perfect system.

6. Because it is an absolutely perfect system, it cannot be an infinite machine, but must be an infinite organism.

7. Because it is an infinite organism, its life-principle must be an infinite immanent Power, acting everywhere and always by organic means for organic ends, and subordinating every event to its own infinite life, — in other words, it must be infinite Will directed by infinite Wisdom.

8. Because it is an infinite organism, its exient organic end disappears as such, but reappears as infinite Love of itself and infinite Love of the finite.

9. Because it is an infinite organism, its immanent organic end appears as the eternal realization of the Ideal, and therefore as infinite Holiness.

10. Because, as an infinite organism, it thus manifests infinite Wisdom, Power, and Goodness, or thought, feeling, and will in their infinite fulness, and because these three constitute the essential manifestations of personality, it must be conceived as Infinite Person, Absolute Spirit, Creative Source and Eternal Home of the derivative finite personalities which depend upon it, but are no less real than itself.

§ 90. Such appears to me to be the conception of the universe which flows naturally, logically, inevitably, from the philosophized scientific method; and such, therefore, appears to me to be the IDEA OF GOD which is the legitimate outcome of modern

14

science. In truth, it is the scientific and strictly *à posteriori* proof of God's existence, attributes, and character, based solely upon the *data of universal human experience of universal Nature*, as organized into the living process of the scientific method, and upon the strictly legitimate philosophizing of that method. New England Transcendentalism [1] denies on *à priori* grounds the possibility of any such proof; but the proof itself now lies before the world, and the world will judge its conclusiveness.

§ 91. The further question, whether this idea of God is Pantheism, is a question of the proper definition of the word, and of far less significance. A score of years ago I named and promulgated this essential idea as SCIENTIFIC THEISM, and I still judge that to be the most appropriate designation of it. If all forms of Monism are necessarily deemed

---

[1] "It is my belief that reason in its original capacity and function has no knowledge of spiritual truth, not even of the first and fundamental truth of religion, the being of God. . . . I deny the ability of the human intellect to construct that ladder, whose foot being grounded in irrefragable axiom, and its steps all laid in dialectic continuity, the topmost round thereof shall lift the climbing intellect into vision of the Godhead. Between the last truth which the human intellect can reach by legitimate induction and the being of God there will ever lie — 'deserts of vast eternity.' Not by that process did any soul yet arrive at that transcendent truth; not from beneath, but from above, — not by intellectual escalade, but by heavenly condescension, — comes the idea of God, even by the condescending Word," etc. (F. H. Hedge, *Reason in Religion*, p. 208, Boston, 1865.) Dr. Hedge's distrust and fear of the understanding, or "human intellect," which is shared by most of the Transcendentalists, arises from defective comprehension of the spirit, tendency, and immanent philosophical creativeness of the scientific method.

Pantheism, on the ground that Pantheism must include all systems of thought which rest on the principle of one sole substance, then Scientific Theism must be conceded to be Pantheism; for it certainly holds that the All is God and God the All, — that the Dualism which posits Spirit and Matter as two incomprehensibly related substances, eternally alien to each other and mutually hostile in their essential nature, is a defective intellectual synthesis of the facts, and therefore greatly inferior to the Monism which posits the absolute unity of substance and absolute unity of relational constitution in one organic universe *per se,* and which conceives God, the Infinite Subject, as eternally thinking, objectifying, and revealing himself in Nature, the Infinite Object. Dualism is inevitably driven to Deism, with its clumsy makeshift of creation *ex nihilo;* and Deism is the only form of the mechanical theory of evolution which does not flatly contradict the mechanical concept. Abundant reasons have already been given why the "monistic" mechanical theory should be rejected; but whatever cogency they may have tells with equal force against Dualism itself, except in the one point of teleology.

§ 92. If, on the other hand, Pantheism is the denial of all real personality, whether finite or infinite, then, most emphatically, Scientific Theism is *not* Pantheism, but its diametrical opposite. Teleology is the very essence of purely spiritual personality; it presupposes thought, feeling, and will;

it is the decisive battle-ground between the personal and impersonal conceptions of the universe. There is no such thing as unconscious teleology; if it is not conscious in the finite organism, as of course it is not in the organic structure as distinguished from the organic consciousness and action, then it must be conscious in the infinite organism which creates the finite. Ends and means are inconceivable and impossible, except as ideal or subjective relational systems which the creative understanding absolutely produces, and which the will reproduces in Nature as real or objective relational systems; hence the recognition of Teleology in Nature is necessarily the recognition of purely spiritual Personality in God. Yet Teleology, say what one will, cannot be escaped by any device in the comprehension of Nature; it is either openly confessed in, or else surreptitiously introduced into, all philosophical systems of evolution, as has been instanced above in the systems of Haeckel and Spencer. Teleology conjoined with Dualism, however, yields only the most awkward and artificial form of the mechanical theory — that of Deism, or the theory of an external creator, creation *ex nihilo*, and meaningless "second causes;" while Teleology conjoined with Monism yields the organic theory of evolution or Scientific Theism, which includes only so much of Pantheism as is really true and has appeared in every deeply religious philosophy since the very birth of human thought.

§ 93. For every deeply religious philosophy must hold fast, at the same time, the two great principles of the Transcendence and the Immanence of God; and that of his Immanence, thought down to its foundation, is Monism. If God is not conceived as transcendent, he is confounded with matter, as in Hylozoism, Materialism, or Material Pantheism. But, if he is not conceived as immanent, he is banished from his own universe as a Creator *ex nihilo* and mere Infinite Mechanic. Scientific Theism conceives him as immanent in the universe so far as it is known, and transcendent in the universe so far as it remains unknown, — immanent, that is, in the world of human experience, and transcendent in the world which lies beyond human experience. This is the only legitimate or philosophical meaning of the word transcendent; for God is still conceived as immanent alone, and in no sense transcendent, in the infinite universe *per se.* Hence the merely subjective distinction of the Transcendence and Immanence of God perfectly corresponds with that of the "Known" and the "Unknown," as absolutely one in Real Being; God is "Known" as the Immanent, and "Unknown" as the Transcendent; but he is absolutely knowable as both the Immanent and the Transcendent. It is really denial of him to confound him with the "Unknowable" or Unintelligible — that is, the Non-Existent. Scientific Theism does not insult and outrage the human mind by calling upon it to worship what it cannot possibly under-

stand — an unreal quantity, a surd, a square root of
minus one, an "Unknowable Reality" which is only
a synonym for Impossible Reality or Absolute Unreality; for that is the quintessence of superstition.
But it gives an idea of God which not only satisfies
the demands of the human intellect, but no less
those of the human heart.

§ 94. In vain will the soul of man strive to worship, to venerate, to love, that which has no intelligible being: the clear idea must precede the vivid
and deep and strong emotion, just as necessarily as
the fountain-head must precede the beautiful river
with its glory of smiling banks. So long as man is
finite, so long indeed will the Mysterious, the Transcendent, the Unknown abide, as the infinite Beyond
to which the finite cannot reach; and the presence
of this ever-abiding Mystery perpetually excites
those sentiments of sublimity and awe which are
indeed the unfailing concomitant of all true worship. But every sentiment of true worship is absolutely extinguished in the intelligent mind where no
clear idea is presented — where no luminous thought
shoots its radiance into the fathomless abyss of
Being, but where all is black with impenetrable
darkness. If the glorious thought of a universe in
which God is at once the Self-Manifesting and the
Self-Manifested, the Self-Revealing and the Self-
Revealed, — a universe in which the adoring Kepler
might well exclaim in awe unspeakable, "O God, I
think Thy thoughts after Thee!" — a universe which

is the eternally objectified Divine Idea, illumining
the human intellect, inspiring the human conscience,
warming the human heart, — if, I say, this glorious
thought begotten of science has no power to stir the
depths of the human soul and lift it up to the sublimest heights of worship and self-consecration to the
service of the Most High, then religion is dead indeed,
and the light of the universe is gone out forever.
But, if this thought of God, the reflected glory of its
divine source, has, as in truth it has, such a divine
force and energy in itself as to soothe the woes of
life, and dull the pangs of sorrow, and minister new
strength to the soul faltering in the path of painful
duty, then religion is not dead, but sleeping, and
will yet rise from its bier at the commanding word
of SCIENCE.

§ 95. Ralph Waldo Emerson, whose great memory
hovers like a benediction over the heads of this
mighty and happy people, uttered, in one of the
latest, if not the very latest, of his public addresses
(and it was my signal privilege to listen to it), this
dignified lament over one of the immediate, yet I
believe transient, effects of the spread of the scientific
spirit in our day : —

"In consequence of this revolution in opinion, it
appears, for the time, as the misfortune of the period
that the cultivated mind has not the happiness and
dignity of the religious sentiment. We are born
too late for the old, and too early for the new, faith.
I see in those classes and those persons in whom I

am accustomed to look for tendency and progress, for what is most positive and most rich in human nature, and who contain the activity of to-day and the assurance of to-morrow, — I see in them character, but scepticism; a clear enough perception of the inadequacy of the popular religious statement to the wants of their heart and intellect, and explicit declarations of this fact. They have insight and truthfulness; they will not mask their convictions; they hate cant; but more than this I do not readily find. The gracious motions of the soul — piety, adoration — I do not find. Scorn of hypocrisy, pride of personal character, elegance of taste and of manners and of pursuit, a boundless ambition of the intellect, willingness to sacrifice personal interests for the integrity of the character, — all these they have; but that religious submission and abandonment which give man a new element and being, and make him sublime, — it is not in churches, it is not in houses. I see movement, I hear aspirations, but I see not how the great God prepares to satisfy the heart in the new order of things."

§ 96. The great seer saw not deeply enough into the recesses of this new scientific spirit; the great prophet of New England Transcendentalism read not deeply enough that mighty *striving after truth* which is born of the scientific method, and in turn bears fruit in the bewildering scientific discoveries of this new time. He saw not the slow and obscure beginnings of a new form of faith, sprung not from

the " ecstatic intuition " of Transcendentalism, but
from a closer contact of the human intellect with
the real universe than was ever possible before, —
heralded, not by the earthquake and the wind of the
great discoveries themselves, but by the "still, small
voice" of their creator, the Scientific Method, which
only those can hear who are patient enough to pon-
der, to meditate, and to muse. If I have rightly
divined the inner character, spirit, and tendency of
this philosophy fated to be, it will not only "satisfy
the heart in the new order of things," but also (condition antecedent to this heart-satisfaction [1]) satisfy
the head as well. For the head has been too long
sacrificed to the heart in religion; and the result
to-day is the satisfaction of neither. Scientific
Theism is more than a philosophy: it is a religion,
it is a gospel, it is the Faith of the Future, founded
on knowledge rather than on blind belief, — a faith
in which head and heart will be no more arrayed
against each other in irreconcilable feud, as the
world beholds them now, but will kneel in worship
side by side at the same altar, dedicated, not to the

---

[1] Dante (*Paradiso*, XXVIII. 106–111) beautifully expresses this
thought that the *vision* of Divine Truth must precede the *love* of it,
and constitute the foundation of beatitude : —

> " E déi saver che tutti hanno diletto,
>   Quanto la sua veduta si profonda
>   Nel Vero, in che si queta ogn' intelletto.
> Quinci si può veder come si fonda
>   L' esser beato nell' atto che vede,
>   Non in quel ch' ama, che poscia seconda."

" Unknown God," still less to the " Unknowable God," but to the KNOWN GOD whose revealing prophet is SCIENCE.

For the idea of God which science is slowly, nay, unconsciously, creating is that of no metaphysical abstraction spun out of the cobwebs of idealistic speculation, but rather that of the immanent, organific, and supremely spiritual Infinite Life, revealing itself visibly in Nature, and, above all, invisibly in Nature's sublimest product — human nature and the human soul. Scientific Theism utters in intelligible speech the very heart, the Infinite Heart, of the universe itself, and speaks with resistless persuasion to the heart of all who can comprehend it. He who can firmly grasp the torch of this self-luminous Knowledge of God possesses an " Inner Light " beside which all other lights are wandering wills-o'-the-wisp, and knows himself to be in absolute security, come what may, so long as he walks the paths of destiny by the clear and steady radiance it sheds, and lifts up his soul in secret loyalty and adoration to Him from whose infinite being all human knowledge itself is a shining ray. With all reverence and tenderness for the illustrious dead be it spoken: I *do* " see how the great God prepares to satisfy the heart in the new order of things." For Scientific Theism is the PHILOSOPHY OF FREE RELIGION and the RELIGION OF FREE PHILOSOPHY.

Art Thou the Life?
To Thee, then, do I owe each beat and breath,
And wait thy ordering of my hour of death
          In peace or strife.

Art Thou the Light?
To Thee, then, in the sunshine or the cloud,
Or in my chamber lone or in the crowd,
          I lift my sight.

Art Thou the Truth?
To Thee, then, loved and craved and sought of yore,
I consecrate my manhood o'er and o'er,
          As erst my youth.

Art Thou the Strong?
To Thee, then, though the air be thick with night,
I trust the seeming-unprotected Right,
          And leave the Wrong.

Art Thou the Wise?
To Thee, then, would I bring each useless care,
And bid my soul unsay her idle prayer,
          And hush her cries.

Art Thou the Good?
To Thee, then, with a thirsting heart I turn,
And at Thy fountain stand, and hold my urn,
          As aye I stood.

Forgive the call!
I cannot shut Thee from my sense or soul,
I cannot lose me in the boundless whole —
          For THOU art ALL.